Reviews in Polymer Technology

Edited by

Irving Skeist

Skeist Laboratories, Inc.
Livingston, N. J. 07039

Volume 1

1972

MARCEL DEKKER, INC., New York

MARCEL DEKKER, INC., 95 Madison Avenue, New York, New York 10016

Library of Congress Catalog Card Number 70-150265

ISBN 0-8247-1627-2

Printed in the United States of America

Preface

The polymer industries—plastics, coatings, elastomers, adhesives, fibers—have come to dominate the chemical industry. In the U.S. alone, plastics production is approximately 20 billion pounds per year, fibers nearly 6 billion, and elastomers 2.5 billion. Not only are these industries growing at a rate far greater than Gross National Product, but new techniques for polymerization and new ways of utilizing them are continuously being discovered.

This series will consist of intensive reviews of the state of the art in the manifold branches of polymer technology.

In the first volume, six reviews are presented. Coupling agents, long appreciated as necessary components of fiberglass-reinforced plastics, are now showing themselves to be desirable components of many other multi-phase systems. In less than a decade, powdered polyethylene has become the preferred material for the backing of automotive carpets, and now is emerging as a favored material for the manufacture of large containers and other hollow articles by the sophisticated techniques of rotational molding. Building and construction utilizes one-fourth of all plastic materials, but the potential for PVC and a dozen other leading polymers is huge. The trend toward industrialized manufacture of building modules and components should accelerate the already rapid penetration of plastics.

Photo-cross-linkable polymers are highly sophisticated materials, which are quite rapidly displacing metals for printing plates and other uses in the graphic arts. Cinnamic derivatives, azido derivatives and still newer photosensitizable materials are examined in detail. Organic colorants offer unique properties to fibers and other polymeric products. Finally, the fire retardance of polymeric compositions is of increasing importance, as building codes and other government regulations become stricter.

From this diversity of topics, one begins to appreciate the increasing complexity of polymer usage. Scientists and engineers who are willing to share their expertise with others are cordially invited to contribute to coming volumes in this series.

I. Skeist

Contents

Coupling Agents As Adhesion Promoters*

PATRICK E. CASSIDY and BILL J. YAGER†

TRACOR, Inc.
Austin, Texas 78721

ABSTRACT

An extensive review has been made of the literature from 1950 to the present to include all work in the area of coupling agents and their use toward improving adhesion. The types of agents studied, mechanisms by which they act, substrates, adhesive systems, and theories of adhesion are all considered. By far the most common system is silane agent on a glass substrate, and a covalent siloxane linkage is proposed as the operative mechanism. Phosphorus esters and chromium-acid complexes are also known to be effective, and similar mechanisms are postulated. Other species will adsorb on adherend surfaces to provide an adhesive interlayer. The use of coupling agents as admixtures with adhesives and the use and mechanisms of their behavior on metallic substrates have not been thoroughly studied until recently.

INTRODUCTION

The science of adhesion, although studied for years, is now experiencing serious investigation of the interfacial molecular forces responsible for this phenomenon. Discussions have arisen over the relative effects of chemical bonding and physical wetting, the best methods to obtain the optimum in

---

*Prepared for National Aeronautics and Space Administration, George C. Marshall Space Flight Center, under Contract NAS8-24073.

†Present address: Department of Chemistry, Southwest Texas State University, San Marcos, Texas 78666.

each instance, and the role of adsorbed species and weak interfacial layers. It is not the purpose of this review to discuss the entire science of adhesion but to concentrate on one area of increasing importance: the formation of a primary chemical bond linkage between a nonorganic substrate and an organic adhesive system by the use of a coupling agent.

The fact that coupling agents are considered a principal factor in adhesion states a priori that the primary chemical bond is of considerable importance in adhesives. This is not as readily accepted as it may appear and is certainly not the whole story. Other matters which must be considered are wetting, interlayers, bulk properties, adsorbed species, environmental effects, etc. One should recognize that all of these parameters play important roles which cannot often be clearly separated.

Wetting of an adherend by the adhesive is, of course, necessary for good adhesion as shown many times over. After all, the function of the adhesive is to provide and maintain intimate contact between two solid surfaces. This is accomplished by having a liquid adhesive with a surface tension (γ_{LV}) which is less than the critical surface tension of wetting (γ_C) of the solid surface (346). When this condition exists, the liquid forms a low (zero) contact angle with the substrate which denotes wetting and spreading on the surface. However, it is when one considers the micromechanics of the adhesive process that questions arise as to the contribution from chemical bonding. It appears that primary chemical bonding to the substrate or the resin does indeed make a contribution although it is difficult to quantify.

Coupling agents may be used to form primary bonds to either or both the adherend and the bulk adhesive. The bond to the organic adhesive can occur by virtue of a reactive function on the agent similar to the monomer used or one which is reactive with the monomeric species. That is to say, an agent may contain a terminal epoxy or vinyl moiety which will copolymerize with the organic system it contacts; or the agent may contain a function such as an amine which will cure an epoxy or urethane resin at least in part. Examples of each of these will be discussed below under the specific systems presented. However, with a polyester-glass system, the nonreactive silane (an aminoalkyl silane) was shown to be detrimental to the physical properties. Evidently, it is possible for a release agent effect to occur if primary bonding does not exist between the agent and the organic phase (78a).

Adhesion promoters which do not bond primarily to the resin phase are commonly long-chain fatty acids terminally substituted with p-chlorophenyl groups (225,274). It is postulated for these compounds that the carboxylic acid end groups are used to adsorb to surfaces of glass, metal, and ceramic, while the pendant alkylphenyl groups offer a hydrophobic, resin-soluble phase (10,53,144,180,275). These particular agents also increase the critical surface tension of the substrate to promote wetting (263). However, in some instances, although the agent has the proper reactive moieties, it will decrease γ_C so that the adherend is wet only with difficulty.

There are, of course, two bonding mechanisms by which an agent can attach itself to a surface: simple van der Waals attraction (dipole-dipole) or actual bond formation of a chemical bond by exchange of electrons. The first is evidenced by studies elucidating the relative adherence of substances with varying polarity of functional groups (303). Studies with polymers containing acid and amide functional groups showed that both functions improved peel strength but that the former was superior (188). It was concluded that electron availability contributes more than does dipole moment to interfacial interactions and therefore adhesion. This view corresponds to that which states that certain functions such as piperidine are chemisorbed to certain metallic surfaces such as iron to produce a relatively strong bond effective in adhesives, coatings, or corrosion inhibitors (12). This chemisorption phenomenon has been postulated to take place by virtue of the donation of the unbonded pair of electrons on the nitrogen to the unfilled d orbital of the surface iron atom (12,97b,115b).

Adsorption of simple polar long-chain compounds has been broken into three types: physical, chemisorption, and precipitated adsorption. The latter two types are very selective on metallic substrates. The mechanical properties and solubilities of each of these adsorbed films are distinctive (330).

The formation of a direct primary bond, postulated for many coupling agents in the silane (290), chromium complex (121), and phosphate ester (267) classes, formerly was studied almost exclusively on glass substrates. Recently, however, such studies have been extended to metal surfaces (54b, 304a,b). The commercially available materials are compounds (M—X) which can be hydrolyzed to give reactive MOH functions which subsequently react with the SiOH surface functions by dehydration to form a very stable Si—O—M bond. These materials are generally used as a surface pretreatment prior to bonding and are partially polymerized on the substrate before addition of adhesive. An entirely new surface then presents itself to the adhesive, a surface which is wettable and reactive with the polymer.

Epoxy resins form their own primary bonds to some extent. Free radicals which form during the cure produce chemical bonds with metallic adherends (21,22). In the realm of attaching adsorbable functions to the polymer backbone some limited success has been seen (102b,253). Where one has the bondable functions on the backbone of an applied polymer new considerations are necessary. One is that, in order for the groups to be effective and reach the substrate surface, the conformational energy of the polymer chain must be overcome; that is, the adsorbable functions must become aligned and in doing so possibly distort the normal chain equilibrium conformation.

However, when a polymeric rather than monomeric adsorber is utilized, significant increases in the adhesion of monomolecular layers is realized. This was dramatically shown for corrosion inhibitors of various molecular weights where a degree of polymerization greater than 4 (adsorbable units) gives an increase in corrosion inhibition by a factor of 10,000 (12,97b,115b). The cause of this synergistic effect has been discussed in terms of the equilibrium between each chemisorbed group and the surface. With a multitude of bonded moieties on a polymer backbone, those few which are temporarily nonbonded in equilibrium cause essentially no disruption in over-all attachment of the polymer (343).

Recent work has elucidated the role of the resin in the vicinity of the interface, where, through coupling agents, a modified region of adhesive is produced (157). With no agent and poor adhesion, debonding and failure occur at the interface. However, with covalently bonded agents, failure is essentially cohesive deep in the resin body rather than adhesive at the substrate. (Good adhesive systems are known to fail cohesively.) Further improvement is realized when an additional "inner layer" or "interphase" is introduced (157,198).This intermediate layer has been used on a macroscale by mixing the adhesive and agent in high concentration and applying in a thin coat as a prebonded primer (202). It is postulated through mathematical stress and modulus analyses that a thicker interphase will enhance yield stress. This region of modified resin is an extension of that produced by a monomolecular layer of the coupling agent. This serves to decrease the stress concentrations over a larger distance and to increase bond strength between two drastically different materials.

It is also possible, however, to effect an apparent increase in bond strength without changing interfacial forces. If one plasticizes the bulk adhesive an increase in strength is seen. This results only from absorption of energy by the organic matrix.

The use in composites is an indication of what is seen in filled polymeric systems. As would be expected, as one increases the interaction between filler and organic matrix, the effect of the filler is magnified (217). Stated another way, the ability of the fillers to enhance matrix properties is limited to and dependent upon the state of bonding at the polymer-filler interface (25).

Fillers show an effect similar to that of increasing the cross-links in a system, e.g., an increase of glass transition temperature. This is because the polymer-surface interactions tend to limit the mobility of the polymer molecule much as does cross-linking (180,181,302). Coupling agents then decrease the molecular mobility at the surface even more effectively than simple surfaces and essentially do produce a cross-link at the surface. This was shown by impregnating glass fibers with $TiCl_4$ and placing these in an olefinic monomer, styrene, resulting in initiation of polymerization at the glass surface. The formation of a chain coagulation structure of particles when properly treated (306) is possible. Thus, a polyvinyl chloride filled with TiO_2 which was modified with octadecylamine demonstrated this ultimate effect of fillers.

Aside from initial bond strength improvement there have been two other distinct advantages generally recognized as resulting from coupling agents. These are (1) a lesser dependence on cure schedule for optimum strength, and (2) a resistance to bond strength degradation by aging or moisture. The second and perhaps more serious effect has been borne out experimentally on glass reinforced thermoset and thermoplastic systems (76a,b,229,308) and with metal-polyurethane bonds (304a,b). This phenomenon is attributed by DeLollis to the fact that water is preferentially adsorbed on the substrate with subsequent displacement of any adhesive which is not primarily bonded. However, when a coupling agent is utilized, a primary bond is formed to the adherend surface which is not susceptible to hydrolysis or displacement by water. Thompson and Hill (304a,b) have also noted that, when a silane adhesion promoter is used in a urethane adhesive, the original strength of a water-saturated weakened bond can be regained by drying the sample.

The first effect, dependence of bond strength on cure schedule, was elucidated when lap-shear strength was determined for an epoxy-aluminum system with varying cure schedule (54c). This strong dependence of adhesion on cure schedule was decreased considerably by the use of coupling agents. This means that manufacturing procedures can be less strictly controlled.

Certain generalities are worthy of note from this review: (1) Most of the work on coupling agents, both from a practical and theoretical standpoint, has been done on glass substrates. Recently, however, there has been increasing interest in the effect of coupling agents on metal surfaces (54b,304a,b). (2) Most experiments have been done by pretreating the surface with the coupling agent prior to application of the adhesive. Again, interest is developing in the incorporation of the agent into the adhesive system because of the economic advantage of a single application (54b,304a,b). However, the integral mixture requires not only that the agent migrate to and react at the substrate surface but also that functional groups be similar in reactivity to the resin so that polymerization is realized only when desired.

The following sections discuss in detail all coupling agents known through June, 1969. The three principal types are silanes, chromium complexes, and phosphorus esters which are discussed separately. A final section is a review of many miscellaneous surface agents on which some experiments have been run. A table of comparative data is also given.

References have been selected from over 1100 references on this subject since 1950. The alphabetical listing by authors is supplemented by location of the abstract in Chemical Abstracts. Furthermore, following each reference is a series of numbers indicating subject matter dealt with therein according to the code numbers given below. Also given is a compilation (General Bibliography) of books and review articles on coupling agents and adhesive technology.

I. Adhesive

1. Epoxy
2. Urethane
3. Elastomer
4. Polyester
5. Phenolic
6. Silicone
7. Other

II. Substrate

11. Copper
12. Aluminum
13. Steel
14. Glass
15. Rubber
16. Plastic
17. Other

III. Agent Type

21. Phosphorus
22. Chromium
23. Piperidine
24. Silane
25. Other

IV. Miscellaneous
31. Environmental effects
32. Surface chemistry
33. Surface preparation
34. Fillers
35. Test methods

SILANES

A large number of silanes of general structure $R_{(4-y)}SiX_y$ have been evaluated as coupling agents. In these compounds the R groups are organofunctional groups which can be chosen for a specific purpose. The X groups may be halides, alkoxides, and/or acyloxy groups, all of which are hydrolyzed under the conditions of application to give $Si(OH)_y$ groups. These silanol groups may in turn react with polar surface groups of the substrate by dehydration to form a primary bond. Thus, the molecule is ambifunctional; it contains polar silanol groups capable of adhering to the surface of glass, metals, etc., and a group R tailored to interact with the adhesive resin.

The adherence of the polar hydroxyl groups to the surface of glass is postulated to be through an ether-type (siloxane) linkage between the SiOH groups of the glass and the $Si(OH)_y$ groups of the silane. Each molecule has the possibility of forming y Si—O—Si bonds. Whether the siloxane bonds are all formed with the glass surface (as in a) or with adjoining silane coupling

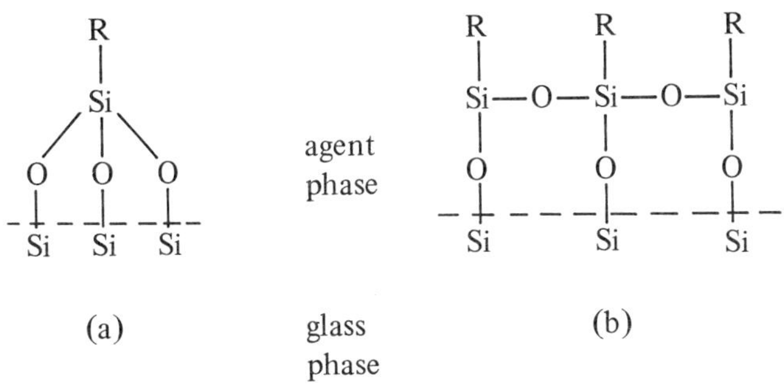

agents (as in b) is still in question. However, it is known that the greater the possible number of siloxane bonds, the greater the stability of the linkage between coupling agents and the surface (343). Further, kinetic studies have shown an increase in bond order between silane and glass in the order of mono-, di-, and trichlorosilane (117). Consequently, hydrolysis is more difficult and water stability is greatly enhanced for glass-resin bonds using as coupling agents silanes of the type $RSiX_3$. While many silanes have been tested (19,156,326,336), it is logical that the most effective and commercially available coupling agents have the general structure $RSiX_3$.

Considerable work has been done in an attempt to determine whether the interaction between substrate and silane is actually a chemical bond as postulated above or simply a physical adsorption. Thermograms of the treatment of quartz powder with CH_3SiCl_3 vapor show a pronounced exothermic effect which is explained in terms of bond formation. Silanes with fewer X groups, $(CH_3)_2SiCl_2$ and $(CH_3)_3SiCl$, show proportionately weaker bonds (11). These results were further substantiated by infrared spectroscopic studies of deuterated glass surfaces which showed weaker bonding on mono- and difunctional silanes as compared to the trifunctional silanes. Only external hydroxyl groups of the glass were found to react and these on a 1:1 basis with $(CH_3)_2SiCl_2$ and $(CH_3)_3SiCl$. The CH_3SiCl_3 reacted more on a 1:2 and 1:3 basis with the glass hydroxyl groups (52,59,117). Isotopically labeled (^{14}C) silanes have also been used to examine the extent of reaction with glass surfaces. Tracer and photomicrographic studies indicate continuous films instead of islands of the coupling agent with covalent bonds at the surface. Failure of the film in boiling water was attributed to failure of the glass substrate and not the glass-silane interface (145,325).

A novel investigation of the electronic interaction between coupling agent and glass substrate involved the determination of charge developed by stripping films of polymers from the glass surface. With untreated glass, the charge on the stripped film was negative, but it was positive when the glass had been treated with silane (167,168). The study is indicative of greater electronic interaction than can be explained by simple absorption of the coupling agent. Chemical methods have also been used with some success to study glass-silane interaction (19,171,337).

Although silanes are useful as coupling agents on substrates other than glass (aluminum, steel, copper, etc.) (54b,304a,b), no definitive work has been published on the mechanism of interaction between the silane and these surfaces. Proposed mechanisms involve either a silanol-type interaction with surface oxide or hydration layers, or chelation of the metal by aminoalkyl groups of the silane.

Silane coupling agents can be divided into three classes according to their reactivity with the organic phase, nonreactive, reactive, and catalytic, and are discussed accordingly below.

Nonreactive silanes have R groups which are alkyl or aryl with no unsaturated or other reactive moieties. With a short alkyl group as in CH_3SiCl_3, a finish which is nonadherent to polar substances such as water and polyurethanes can be imparted to surfaces (19,80,94,102a,103,104,163). However, methyl silanes have been used with excellent results as a primer for silicone rubber bonding to many surfaces (120). Usually longer alkyl groups (84,278), aryl groups (32,78a,226,332), aralkyl groups (221,225), and halogenated derivatives of these groups (122,225,259) are incorporated in the silane for cohesive interaction with the bulk adhesive (116,197,239,278). Such groups are effectively used with nonpolar resins such as polystyrene and other polyvinyl compounds.

Reactive silanes contain a function capable of chemically reacting with the bulk adhesive commonly as a co-monomer. Reactive functions commonly employed are vinyl groups (9,61,111,114), epoxides (31,33,34), mercaptans (105), alcohols (106,175), phenols (197), acids (60), esters (82,85), and combinations of these functions (153,281,295). The choice of reactive group depends on the adhesive to be used and the type of interaction possible. Obviously, compatibility of the R groups and the resin is necessary even in the case of the nonreactive silanes. To provide compatibility, polarity of the resin should match the polarity of the functional group to the extent of matching the degree of hydrogen bonding in each. A summary of the various reactions which may occur between resins and functional groups has been prepared by Plueddemann et al. (241):

(1) Condensation reactions are possible between hydroxyl, carboxyl, mercapto, amino, or epoxy groups on silicon with hydroxyl or carboxyl groups of a polyester.

(2) Olefinic hydrocarbons or unsaturated ether and ester groups on silicon may participate in free-radical-induced addition polymerization with an unsaturated polyester and styrene monomer.

(3) Almost all functional groups are capable of reacting with an epoxy resin or its curing agent in an epoxy laminate.

The third class, the catalytic silanes, contain amino groups which can act as catalysts in the polymerization of phenolic, urea, and melamine resins and can act as curing agents for epoxy (17,160,183,294,296,313) and polyurethane (6) resins. Thus, the coupling agent is integrally involved with and bonded to the bulk adhesive. Examples of the amino groups present in commercially available catalytic silanes are $—(CH_2)_3—NH_2$, $—(CH_2)_3—NH—(CH_2)_3—NH_2$ and $—(CH_2)_3—NH—(CH_2)_2—NH—(CH_2)_2—CO_2CH_3$ (9,15,85, 102a,118). In reaction with some adhesive systems, the amino silanes may be classified as simply reactive silanes and not catalytic. As such the amino group is capable of reacting with alkylhalides, acids, anhydrides, and esters in substitution or condensation reactions. (Oddly, the amino silanes give poor adhesion to polyester resins, perhaps because of amine inhibition of the polyester cure (241).)

Silane coupling agents are usually applied to a glass substrate surface as a pretreatment from a dilute aqueous solution. Acid or base is added to the solution to aid in the hydrolysis of the SiX groups to SiOH. The solution is necessarily dilute (0.5-1.0% coupling agent) since optimum coupling effect is achieved with a monomolecular layer rather than a multilayer film of the silane. The surface is then dried, usually at temperatures over 100°C, to promote the condensation reactions and to remove excess water. Vapor phase application of alkyl-alkoxy silanes is accomplished by mixing the silane with moist air for hydrolysis and applying it to nascent glass surfaces while the temperature is held near the decomposition point of the silane (89).

Some integral blending of the coupling agent into the adhesive resin has been tested with encouraging results (6,27,36,54b,87,97a,105,108,115a,123, 158,234c,290-292,304a,b). For example, outstanding improvement of metal bonding has been reported when silane coupling agents were integrally mixed with urethane and epoxy adhesives (54b,304a,b). Such application has the economic advantage of reducing the number of steps required in the adhesive process. However, this advantage is somewhat negated in the usual procedure for integrally mixed adhesive systems because (1) a slower curing process may be needed to allow migration of the coupling agent to the substrate surface and (2) an excess of coupling agent is usually required since apparently all of the coupling agent does not reach the adherent surface.

The preceding discussion has dealt primarily with silane glass systems. More work has been published in this area than about other substrates because of the long-time interest in glass lamination in composites. However, silane coupling agents, as surface pretreatments or as mixtures, are being used increasingly in the bonding of metal substrates with a variety of adhesives (54b,304a,b). The brief discussion below of some of the systems which have been reported will give some indication of the extent of such usage.

First, it should be noted that improved properties other than total strength may be obtained through the use of silane coupling agents with certain adhesive systems. The results of a long-term aging study have shown that more reliable and much improved adhesive properties were achieved when silane coupling agents were used with polyether-based polyurethane adhesive on aluminum. During a three-year period, loss of strength with normal variations in atmospheric humidity has been minimized, and greatly increased strengths at room temperature and +200°F have been observed (304b). Furthermore, the samples utilizing a coupling agent can be dried to reproduce the original strengths.

Copper, treated with aminoalkyl silane compounds, was effectively bonded to silicone elastomers with exceptional bond stability at elevated temperatures (34,234a,b). Considering the substrate and adhesive, the probable mechanism was the chelation of the copper with the amino groups.

Aluminum has been treated with silanes to improve bonding with several adhesives. Bonding to silicone elastomers was enhanced either by amino silanes incorporated into the silicone (49,70,182,248,249) or by methyl silane treatment of the metal surface (97a,120). Vinyl silane treatment of the metal also aided the adhesion of silicone elastomers (138,156). Adhesion of polyvinyl halides to aluminum and steel surfaces was increased by addition of amino silanes (277). Polyurethane bonding of aluminum was greatly increased (particularly at +200°F) by surface treatment of the metal with an amino- or epoxy-containing silane (304a,b). Furthermore, incorporation of amino and epoxy silanes into the polyurethane adhesive has been tested on aluminum with good results (54b, 87, 260, 304a,b). Amino and mercapto silanes have been proved effective as integral parts of adhesive mixtures of epoxy (54b), phenolic, and polyvinyl butyral resins (244).

Steel plates bonded with an epoxy-silane-modified polyurethane adhesive withstood 24 hr in boiling water as compared to only 1 hr with the unmodified adhesive (6). Similarly, the incorporation of an epoxy or amino silane into an epoxy resin increased the seawater resistance of the resin coating on steel (30,86). This serves to point out the contention that the primary benefit of coupling agents may be their effect on aging characteristics rather than on total initial strength of the adhesive (76a,b,304a,b). The proper choice of coupling agents will also allow effective bonding of "contaminated" steel. An epoxy silane treatment of oily or rusty steel increases the lap-shear strength of epoxy bonds over those of clean metal or untreated contaminated metals (85).

Bonding of rubber to metal substrates such as copper, aluminum, or steel was improved by the use of amino silanes either as a pretreatment of the surface or by incorporation into the rubber (182,248). Similarly, amino-silane-treated glass showed better adhesion to synthetic rubber such as neoprene (191). In addition, a "double" treatment of glass with an unsaturated silane followed by an elastomer containing a free-radical curing agent was shown to increase the adhesion to natural or synthetic rubber (317).

Use of coupling agents in bonding of plastic materials usually has been limited to lamination procedures in which the silane was applied to the reinforcing material and the organofunctional group interacted with the resin as it polymerized (191,210,258,282,316,319,323). A study of the structural features of silanes which increase wet and dry flex strength of glass-polybenzimidiazole laminate at high temperature (600°F) has concluded that two conditions are necessary: a stable aromatic bond to silicon and a reactive functional group for bonding to the resin. Thus, cyanophenyl, carboxyphenyl, and bromotolyl groups were adjudged as most effective (238).

Surface treatment of polymeric adherends with coupling agents has been done (124). Impregnation of polyester and polyamide fabrics with glycidyl silanes has been used as a pretreatment for coating the fabrics with silicone rubber prior to polymerization and vulcanization (81). Amino silanes have been used to treat a polyester film used for heat-pressure bonding to polyethylene film to give a highly water-resistant bond (83).

Some substrates to which silanes have been applied must be termed "miscellaneous." A priming of clay surfaces by amino silanes is claimed to increase the adhesion of polyurethane and epoxy coatings and to aid in the water resistance of these coatings (287). Several adhesive mixtures have been formulated which contain various silanes as integral constituents. These adhesives have been effectively applied to a variety of substrates such as paper, brass, wood, and ceramics (57,271).

Much of the preceding discussion has centered on the use of silanes with the widely used adhesives—the epoxides, urethanes, polyesters, etc. The development of new adhesive resins such as polyimides, ionomers, and polysulfones will require the development and evaluation of more silane coupling agents tailored for each of these resins.

CHROMIUM COMPLEXES

Trivalent chromium complexes which include an organic acid anion as one of the complexing species are widely used as coupling agents. The charge on the complex is determined by the coordinating groups which may be anions (usually chloride ions) or neutral molecules (water or alcohols). The structure of such a complex may be pictured as

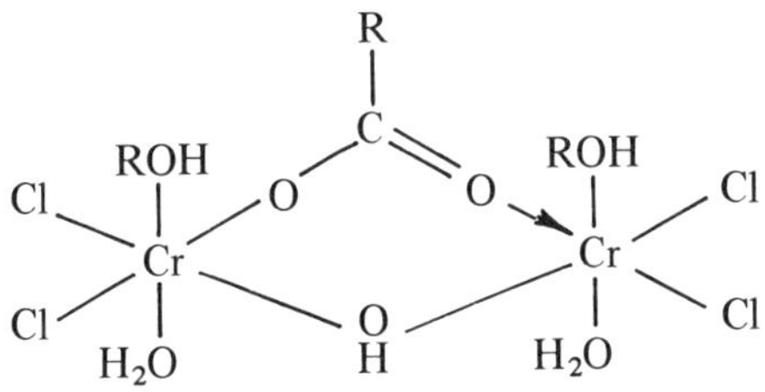

Dilution with water, particularly within pH ranges from 5 to 7, begins hydrolysis in which some chloride ions are replaced by hydroxyl groups as shown below (343):

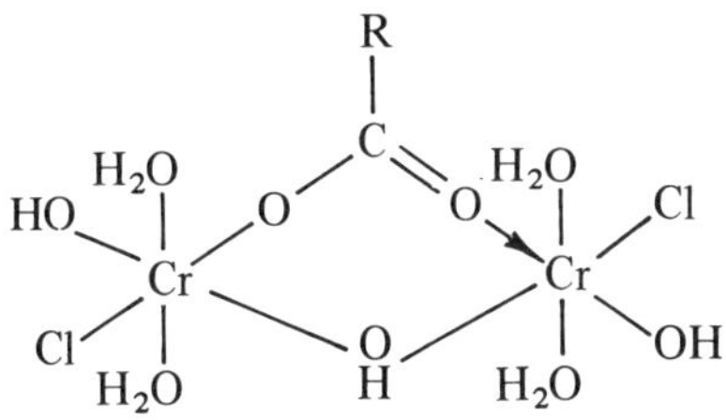

As hydrolysis continues, polymerization begins by the process of olation whereby the complexes are connected by hydroxyl groups:

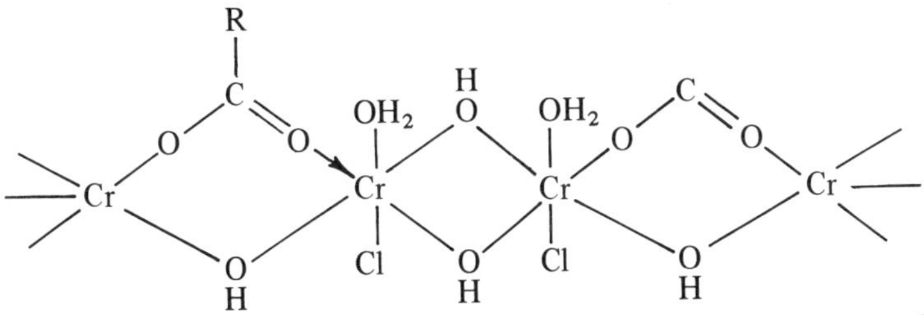

Olation continues with a resulting increase in the size and positive charge of the complex. Absorption of the complex by a substrate surface may occur at any time in the olation process and is thought to occur at negatively charged sites on the surface. Thus, acidic positions such as SiOH in glass will attract the complex if the pH is high enough to encourage ionization to SiO^-. The pH necessary for effective coupling varies with the type of glass owing to variation in ionization of acid groups with glass composition. The bond between a glass surface and the complex might be shown as (121)

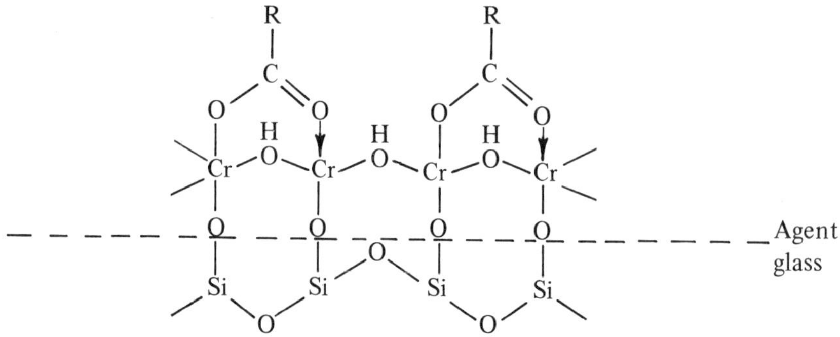

Interaction of the chromium complex with the adhesive is of three types (343):

(1) Coordinate covalent bonding between carboxyl, alcohol, and amine end groups.

(2) Reactions between CrCl or CrOH groups and alcohol end groups to form Cr—O—R linkages with the elimination of either HCl or H_2O.

(3) Copolymerization between the double bonds of methacrylic acid and reactive vinyl-type double bonds in the resin.

Interaction of types (1) and (3) depends largely on the type of acid used in making the complex. The acids can be broadly classified as having reactive or nonreactive R groups. Reactive R groups are unsaturated (for copolymerization) or contain functional groups capable of participating in hydrogen bonding. In addition to methacrylic, acrylic (337), phthalic (3), and succinic acids (230) have been used as "reactive" complexing acids.

"Nonreactive" acids are exemplified by stearic acid which is used principally to impart a hydrophobic finish to surfaces (166) and does not aid adhesion of glass to polyester resins (121). However, stearic acid-chromium complexes are useful in treating $CaCO_3$ filler for such resins. Complexes of several carboxylic acids, saturated and unsaturated, mono- and dibasic, were evaluated for effectiveness in the surface treatment of $CaCO_3$ filler for absorption of polyester resins, and caproic acid was found to be superior (2).

Chromium complexes have been used mainly to treat glass surfaces with only limited application to other substrates. Application is made by dipping the glass surface in a dilute aqueous solution of the acid-chromic chloride complex whose pH is maintained near 6. Aging of the complex to allow olation and polymerization has been shown to greatly improve performance as a coupling agent. The surface is then dried at ~150°C prior to the addition of the adhesive polymer. (An exception in which no adhesive is used is a novel method of preparing inner reflecting glass spheres by dropping hot spheres into a cold solution of methacrylato-chromic chloride. The complex is the bonding agent which holds the fractured spheres together (339).

Other substrates to which chromium complexes have been applied include cellulose (100), mineral fibers (78b), rubber (69), leather, and textiles (3,166). Cellulose and mineral fibers were treated to increase their adherence to polyolefin films. Rubber was given increased moisture resistance by treatment prior to vulcanization. Hydrophobic finishes were imparted to leather, paper, and other textiles by treatment with a stearic or oleic acid complex. A related process involves addition of chromium salts to polyvinyl alcohol and polyacrylic acid to increase their elasticity, water resistance, and film-forming capacity (230).

Chromium complexes have been used as adhesion promoters with a variety of adhesives. Epoxy resins have been widely used in joining treated glass surfaces (109,131,227), in bonding treated mica in electrical insulators (149), and in moldings with treated SiO_2 filler (252).

Polyester adhesives have been used to bond treated glass surfaces (1,126, 129,216,275,334). Chromium-complex-treated sand filler has been molded with polyester resin to produce stain-resistant cast-resin surfaces (51). Combinations of polyester and epoxy resins have been used on treated glass laminates (8) and in molding treated mica filler (149).

Phenolic resins have been used to laminate glass fibers which had been previously treated with methacrylato-chromic chloride (44,129). Phenolic, epoxy, polyester, and organo silicon resins were tested alone and in combinations for adhesion to treated glass surfaces (109) where epoxy resins were found to be superior.

Several investigations (27,126,129,131,275) have compared the efficiency of chromium complexes and silanes in promoting adhesion of glass substrates to various adhesives. The silanes were always found to be superior, particularly with polyester adhesives.

While chromium complex coupling agents increase dry bond strengths significantly, their real value is in increasing wet strength or moisture resistance of bonds. For instance, dry strength of glass-epoxy laminate was increased by 30% while wet strength was increased by 100% over that of the untreated substrate (67). However, in one case (27) methacrylic acid-chromium-complex-treated glass was found to have poorer resistance to water absorption than untreated glass when bonded with polyester. These poor results are attributed to experimental and handling techniques since improvement in wet strength is obtained if the Cl:Cr mole ratio is between 0.1 and 1.2 (optimum 0.6 and 0.7) when the methacrylato-chromium complex is prepared. The Cl:Cr ratio is adjusted by varying the amount of CrO_3 and HCl added to methacrylic acid (307).

PHOSPHORUS-CONTAINING COMPOUNDS

Some phosphorus compounds are known to behave as coupling agents. Esters of phosphoric, phosphonic, and phosphorous acids react with epoxy resins and decrease their viscosities (133). Wollastonite, when treated with H_3PO_3 before addition to a polyester containing tritolyl phosphate (TTP), contributes significantly to the flexural strength of the filled material (206). In place of TTP one can utilize phosphate esters substituted with allylic, vinyl, ester, and aminoalkyl groups. It is postulated that the mineral filler is bonded to the polymeric matrix by virtue of a phosphorus linkage.

A considerable amount of definitive work has been done with these phosphorus agents at the Naval Applied Science Laboratory by Schrader, Lerner, D'Oria, and Deutsch (263-268). Triethyl phosphate was found to adsorb irreversibly on E glass surfaces in a high vacuum system. Desorption-decomposition experiments at high temperatures indicated the original adsorption mechanisms to be

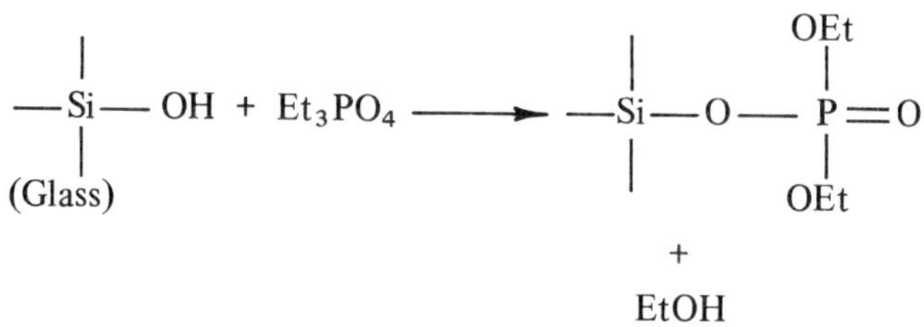

(1)

and

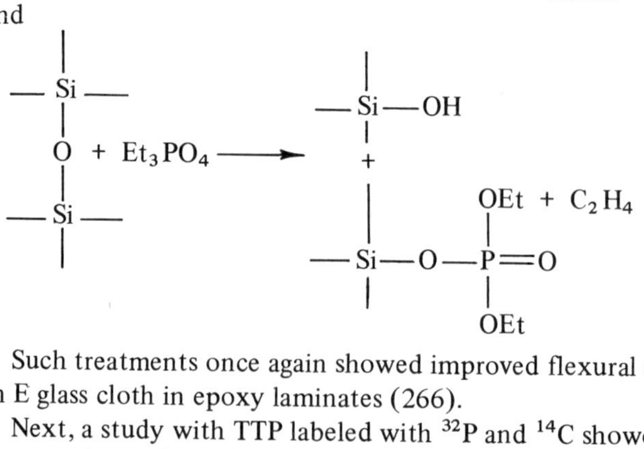

(2)

Such treatments once again showed improved flexural strengths when used on E glass cloth in epoxy laminates (266).

Next, a study with TTP labeled with ^{32}P and ^{14}C showed two reactions to occur at the surface: (1) the formation of a primary chemical bond, Si—O—P, between TTP and the glass surface; and (2) hydrolysis of TTP by adsorbed water to give cresol and H_3PO_4 (265), which contributes to the

DeLollis desorption mechanism (76a). Further tracer studies showed a three-component film present upon adsorption of diethyl phosphite (DEP): (1) a water-soluble material at 25°C which is physically adsorbed, (2) a species chemisorbed through a single Si—O—P bond which is insoluble in water at 25°C, and (3) a water-insoluble portion at 100°C which is bonded through multiple Si—O—P bonds. The physically adsorbed layer is believed to participate in increasing the dry strength of epoxy laminates (264). The general method for applying DEP to glass is to place the substrate in a 1% aqueous solution of DEP, air dry, and heat at 150°C for 10 min (263).

Similar multicomponent films were found with ^{14}C-labeled (γ-aminopropyl)-triethoxy silane (A-1100) (267). An increase in the temperature of adsorption of these agents increases the stability (irreversibility) of the layer. This effect is because at higher temperature a polymeric monolayer is formed with multiple sites of attachment to the surface (268). Thus, when one surface bond is reversibly broken by hydrolysis it has time to reform before other functions are hydrolyzed. This is the same synergistic effect experienced by Hackerman and co-workers (11,97b,115b) when studying polymeric piperidinyl corrosion inhibitors. They noticed an increase in corrosion inhibition (by chemisorption) by four orders of magnitude when the degree of polymerization of the adsorbed species was greater than four.

Organophosphorus compounds have also shown utility in increasing the adhesion of polymeric substrates to metals and other organic adhesive systems. Thus, radiation grafting of $(ClCH_2CH_2O)_2$ $P(O)CH{:}CH_2$ to polypropylene improves its adhesion to metals with only a small change in bulk properties (179). This method also allows the study of interfacial effects independent of the bulk properties. This same study claims that interfacial effects act only up to a critical minimum value above which the bulk mechanical properties of the adhesive are the sole determinants of total strength. This contributes to the observation of the high strengths obtained when one achieves cohesive failure.

Similarly, ethylene-vinyl acetate copolymers show increased adhesion to metals when they are peroxide cross-linked in the presence of trialkyl phosphate, where the alkyl groups are incorporated into the backbone leaving pendant phosphate moieties (67).

Thermoplastic solution adhesives containing TTP are used for bonding polyvinyl chloride (PVC) to aluminum, steel to zinc, and for acrylonitrile-butadiene-styrene (ABS) to aluminum (5).

Even fluorinated polyolefins show adhesive improvements on treatment with phosphites. Dipping poly(trifluorochloroethylene) in dibutyl phosphite at 150°C allows it to be bonded to an acrylate-acrylic acid copolymer sheet under pressure (68a). The mechanism in this case may be different than with metallic substrates in that here a transesterification of the surface phosphite may be the bonding mode.

MISCELLANEOUS AGENTS

Acids

Several different types of carboxylic acids have been used as coupling agents on metals, glass, and other substrates. The polar carboxyl group has the capability of coordinating or reacting with polar surface groups such as the SiOH of glass or the AlOH of aluminum in much the same fashion as do chromium complexes or phosphorous compounds:

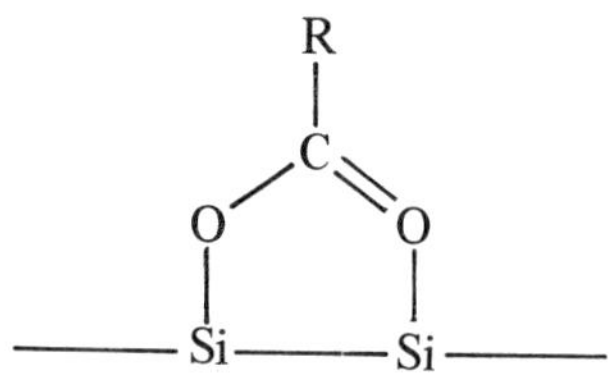

Probably the stability of this complex is reduced because of the fewer coordinating positions of the carboxyl group relative to the chromium or phosphorous compounds. The presence of an ortho-hydroxyl or thiol group in the more effective salicylic (137,177) and thiosalicylic acids (231) offers the possibility of a third coordinating position which would increase stability. Salicylic acid has been incorporated into a phenolic resin to form an adhesive which is effective on a wide variety of substrates (177). Salicylic acid has also been used as a pretreatment of glass for resin adherence (137) and thiosalicylic acid as a pretreatment of cord fabric to increase rubber adhesion (231).

The carbon chain, R, has been varied considerably in the evaluation of coupling effectiveness of acids. The same concepts apply to the R groups of the acids as to those of the silanes and chromium complexes, and the group may be reactive or nonreactive toward the adhesive resin. The 17-carbon saturated R-group of stearic acid exemplifies the nonreactive types. It has been shown to increase the bond strength of metals bonded with polyvinyl-formal-phenol resin by 5-500% if applied as a 0.02-0.06% part of the resin, whereas higher concentrations weakened the bond (215).

Other investigators have shown the value of a monomolecular layer of the acid applied to the substrate. Application has been made either by carefully dipping the substrate in water which is covered by a monolayer of stearic acid (10) or by electrophoresis (255). Stearic acid has been shown to greatly improve the moisture resistance of polyethylene bonds between treated aluminum plates (10,62,324a).

The extraordinary effectiveness of the p-chlorophenyl group as the terminal substituent of R groups in increasing adhesion is attributed to the relatively high energy chlorophenyl outer surface exposed to the adhesive resin, thus producing a surface which is easily wet by the adhesive. The p-chlorophenyl moiety has been used as a terminal substituent on 12-, 14-, 18-, and 20-carbon carboxylic acids and as an ethyl silane substituent (18,225,273).

Acids with reactive R groups, such as maleimidopropionic or succinimidopropionic, have been used as coupling agents on several metals and alloys for epoxy bonding. Aluminum bonded in this manner showed good resistance to boiling water for 48 hr (274). Incorporation of unsaturated acids into the backbone of the polymer chain has been shown to be an effective way of increasing the adherence of the polymer to polar surfaces. The concentration of acid must be kept low to prevent association between the carboxyl groups which reduces surface adherence (53,143).

Titanates

Increasingly of late, tetra esters of ortho-titanic acid have been mentioned as coupling agents for glass, metals, and polymers. While no published mechanism for their action was found, it can be postulated that some of the ester linkages are hydrolyzed and coordination or condensation occurs between the resulting hydroxyl groups and surface groups. Furthermore, polymerization between adjacent titanate groups is likely, considering the effectiveness of poly(dialkyl titanates) as coupling agents (133). Thus the interface between a polar surface and a titanate coupling agent might be pictured as

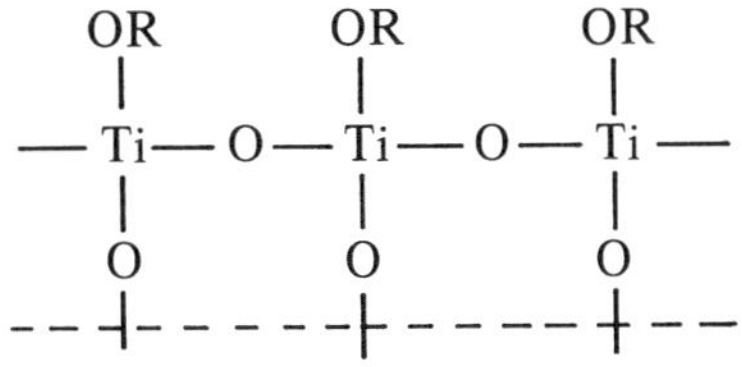

The choice of alkoxy groups on the titanate determines the rate of hydrolysis, which has been reported to be the critical factor in proper surface preparation (112). Butyl and isopropyl groups are the most widely used.

Alkyl titanates are also effective coupling agents for polyethylenes (112, 113,150) and interaction with such a nonpolar surface is more difficult to envision. In any event, at least one alkoxy or hydroxy group is available for interaction with the adhesive resin by dipolar attraction or reaction. In addition to polyolefins, polyesters also benefit from titanate treatment prior to coating with polyethylene (37,43).

Aluminum, copper, and steel have been treated with a "primer" composed of an amino silane mixed with tetraisopropyl titanate before bonding with a silicone adhesive (95). Copper showed the greatest improvement in bond strength. A similar mixture of silane-titanate has been incorporated into polymers (vinyl, epoxy, or ester) for coating glass fibers (254). Titanium chelates such as titanium acetyl acetonate were also used in place of the tetra ester. Use of the chelates on glass surfaces has been investigated by equilibrium studies, and the possibility has been raised of acidic surface sites catalyzing both the chelate hydrolysis and subsequent condensation with the surface silanol groups (344).

Amines

Amines have been used to a limited extent as coupling agents. Quaternary ammonium salts, which exhibit a positively charged nitrogen to bond with nucleophilic surface positions, have been used to increase adhesion of resins to glass (39). A salt having at least one long-chain hydrocarbon radical is used to pretreat the glass surface.

Polyester films and fabrics show excellent adhesion to rubber if they are treated prior to vulcanization with a diamine such as 1,2-diaminoethane. Tear strength was increased fortyfold over untreated polyesters (316).

The effectiveness of certain amines in preventing aqueous corrosion of steel surfaces has promoted interest in their use as coupling agents (342). Amines have been shown to be chemisorbed in a monolayer on the steel by virtue of the unbonded pair of electrons on the nitrogen and the vacant d shell of iron (12,97b,115b). Thus, 4-ethylpiperidine, which is particularly well adsorbed because of favorable polar and steric factors, bonds through the nitrogen atom and thus exposes the ethyl group for interaction with an adhesive resin. Further investigation is needed to determine optimum concentrations, substituent groups, substrates and adhesives to utilize piperidine coupling, although 4-

aminomethylpiperidine has been shown to be beneficial in steel-epoxy systems (54b).

COMMERCIALLY AVAILABLE AGENTS

Table 1 is a compilation of commercially available coupling agents and their suppliers along with an elucidation of their structures. The list is divided along the same major lines as the entire paper, e.g., silanes, chromium complexes, and phosphorous compounds with appropriate subdivisions.

Table 1. Commercially Available Coupling Agents

I. Silanes R—SiX_3 (X = Hydrolyzable Group)
   A. Saturated alkyl and haloalkyl R group
      1. Methyltrichloro silane, CH_3 $SiCl_3$ (General Electric S.C.-3100 and Stauffer SWS-442)
      2. Ethyltriethoxy silane, CH_3—CH_2—$Si(OEt)_3$ (General Electric SC-3735)
      3. γ-Chloropropyltrimethoxy silane, Cl—CH_2—CH_2—CH_2—$Si(OCH_3)_3$ (Dow Corning Z-8-0999)
      4. Dimethyldichloro silane, $(CH_3)_2SiCl_2$ (Stauffer SWS-441)
   B. Aryl and haloaryl R group
      1. Phenyltrimethoxy silane, C_6H_5—$Si(OCH_3)_3$ (Dow Corning 6071)
      2. p-Chlorophenyltrichloro silane, Cl—C_6H_4—$SiCl_3$ (K&K 13181)
   C. Unsaturated alkyl R group
      1. Vinyltriethoxy silane, $CH_2{=}CH$—$Si(OEt)_3$ (Union Carbide A-151 and Dow Corning Z-6075)
      2. γ-Methacryloxypropyltrimethoxy silane, $CH_2{=}CH{-}C({=}O){-}O{-}CH_2CH_2CH_2{-}Si(OCH_3)_3$ (Union Carbide A-174 and Dow Corning Z-6030)

Table 1 (continued)

D. Epoxy R groups

1. γ-Glycidoxypropyltrimethoxy silane, $H_2C\overset{O}{—}CH—CH_2O—CH_2CH_2CH_2—Si(OCH_3)_3$ (Dow Corning A-6040 and Union Carbide A-187)
2. β-(3,4-Epoxycyclohexyl)-ethyltrimethoxy silane, 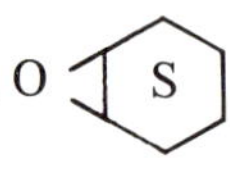 —$CH_2CH_2—Si(OCH_3)_3$ (Union Carbide A-186)

E. Amino R group

1. γ-Aminopropyltriethoxy silane, $H_2N—CH_2CH_2CH_2—Si(OEt)_3$ (Union Carbide A-1100)
2. N-β-(Aminoethyl)-γ-aminopropyltrimethoxy silane, $H_2N—CH_2CH_2—NH—CH_2CH_2CH_2—Si(OCH_3)_3$ (Union Carbide A-1120 and Dow Corning Z-6020)
3. $CH_3—O—C(=O)—CH_2CH_2—NH—CH_2CH_2—NH—CH_2—CH_2—CH_2—Si(OCH_3)_3$ (Dow Corning C-600)

F. Mercapto R group

γ-Mercaptopropyltrimethoxy silane, $HS—CH_2CH_2—CH_2—Si(OMe)_3$ (Union Carbide A-189)

II. Chromium Complexes (Acid Functions Given)

A. Methacrylo, $H_2C{=}C(CH_3)—CO_2—$ (Dupont – Volan)

B. Stearo, $CH_3(CH_2)_{16}—CO_2—$ (Valchem – Valchrome 5015)

III. Tetraalkyl Titanates $(R—O)_4\,Ti$ (R—O Groups Given)

A. Isopropyl, $(CH_3)_2\,CH—O—$ (Stauffer)

B. n-Butyl, $CH_3—(CH_2)_3—O—$ (Stauffer)

C. 2-ethylhexyl, $CH_3—(CH_2)_3—CH(CH_2—CH_3)—CH_2—O—$ (Stauffer)

D. Cresyl, $CH_3—C_6H_4—O—$ (Stauffer)

IV. Phosphorus Compounds (R Groups Given)

A. Phosphates

1. Monoalkyl acids $R—O—P(=O)—(OH)_2$
   a. Butyl (Stauffer and Mobil)

Table 1 (continued)

b. Ethyl (Stauffer and Mobil)
c. Isoamyl (Stauffer and Mobil)

B. Phosphonates $R—\overset{\overset{\large O}{\|}}{P}—(OR)_2$
  1. Diphenylphenyl (Stauffer)
  2. Dimethylmethyl (Mobil)
  3. Diethylethyl (Mobil)
  4. Bis-(2-ethylhexyl)-2-ethylhexyl (Mobil)
  5. Dibutylbutyl (Mobil)

C. Phosphites $(R—O)_2P—OH$
  1. Dimethyl (Stauffer)
  2. Diethyl (Stauffer)
  3. Diisopropyl (Stauffer)
  4. Dioctyl (Stauffer)
  5. Weslink E (Weston Chemical Co.) (exact structure unavailable)

It should be noted that several of these compounds, particularly in the phosphorus section, are not marketed as coupling agents; however, since this review of the literature has found evidence that these types of compounds can be effective coupling agents, they are included in the compilation. This collection is not meant to be complete with respect to commercial sources but presents a comprehensive list of types available.

EVALUATION DATA

Most of the preceding discussion of the effectiveness of coupling agents in increasing dry and wet adhesive strength has been in qualitative terms. Some quantitative tests have been conducted, but direct comparison between reported values is difficult because conditions were varied considerably. However, some of the reported data were selected for presentation in Table 2 to convey some insight into the adhesion-enhancing ability of the coupling agents. Items were chosen to represent as wide a variety of coupling agents, adhesives, substrates, and test conditions as possible from the available data. Bond improvement is reported as the ratio of bond strength with coupling agent to bond strength without coupling agent.

Table 2. Quantitative Evaluation of Coupling Agent Effectiveness

Type	R group	Adhesive	Substrate	Test type	Test conditions	Bond improvement (treated/untreated)	Reference
Silanes $R—SiX_3$	α-Amino propyl	Polyolefin	Glass-rubber	Peel strength	Dry	7.8	(65)
	α-amino propyl	Epoxy-phenolic	Glass laminate	Tensile strength	Dry	1.3	(110)
					2 hr boiling water	1.6	
	Vinyl	Polyester	Glass laminate	Flex strength	Dry	1.3	(162)
					2 hr boiling water	3.0	
	Methyl methacrylate	Polyester	Glass laminate	Flex strength	Dry	1.4	(162)
					2 hr boiling water	4.2	
	3-Glycidoxy propyl	Polyester	Glass laminate	Adhesive strength	Dry	1.4	(228)
	3,4-Epoxy cyclohexyl-ethyl	Polyester	Glass laminate	Adhesive strength	Dry	1.2	(228)
	p-Cyanophenyl	Polybenzimidiazole	Glass laminate	Flex strength	Dry	1.4	(240)
					2 hr boiling water	1.4	

	Mercapo propyl	Phenolic-acrylonitrile	Aluminum	Lap shear	73°F	2.1	(244)
					250°F	3.3	(244)
	Amino propyl	Phenolic-acrylonitrile	Aluminum	Lap shear	73°F	1.6	(244)
					250°F	2.3	(244)
	3-Glycidoxyl propyl	Polyurethane	Aluminum	Lap shear	75°F	1.7	(304a,b)
					−300°F	1.3	(304a,b)
					+200°F	4.4	(304a,b)
Chromium complex	Methacrylic acid	Polyester	Glass laminate	Flex strength	Dry 3 hr boiling water	1.3 1.7	(121)
	Trihydroxybenzoic	Epoxy	Glass laminate	Flex strength	Dry 100 hr boiling water	1.2 2.2	(121)
Carboxylic Acid $R—CO_2H$	Salicylic	Chloroprene-Neoprene copolymer	Canvas	Peel strength	100°F	1.6	(177)
					200°F	1.7	(274)
	Succinimido-acetic	Epoxy	Aluminum	Shear strength	Dry 48 hr boiling water	1.6 (2000/0)	(274)

ACKNOWLEDGMENTS

The authors wish to express their appreciation to the National Aeronautics and Space Administration, George C. Marshall Space Flight Center, Non-Metallic Materials Division, for their financial support and technical assistance on this program, in particular Mr. L. M. Thompson and Dr. W. E. Hill. Gratitude is also due Mr. Gary C. Rolls and Mr. James M. Johnson of TRACOR for the competent aid given throughout the program.

REFERENCES*

(1) J. I. Aber, U.S. Pat. 3,262,809 (July 1966); *Chem. Abstr.,* **65**, 13889g (1966); 4, 7, 14, 22.

(2) T. Akita, and N. Kurihara, *Reinforced Plastics* (*Tokyo*), **4** 2-5 (1958); *Chem. Abstr.,* **53**, 1809h (1959); 4, 22, 25, 34.

(3) S. O. Aktiebolaget, British Pat. 788,539 (Jan. 1958); *Chem. Abstr.,* **52**, 13796e (1958); 7, 17, 22.

(4) V. I. Alksne, K. P. Grinevich, and I. E. Yakobson, Foreign Technol. Div., Wright-Patterson AFB, Ohio, Rept. No. FTD-HT-23-682-68, (Nov. 7, 1968); *USGRDR,* **69**(5), 124 (1969); 4, 14, 16, 24, 31.

(5) Aluminum francais, French Pat. 1,192,267 (Oct. 1959); *Chem. Abstr.,* **54**, 26021c (1960); 4, 12, 13, 16, 17, 21.

(6) American Cyanamid Co., Netherlands Appl. Pat. 6,508,268 (Jan. 1966); *Chem. Abstr.,* **64**, 19922d (1966); 2, 13, 24, 31.

(7) American Cyanamid Co., Netherlands Appl. Pat. 6,605,081 (Oct. 1966); *Chem. Abstr.,* **66**, 66168e (1967); 2, 17, 24.

(8) G. D. Andreevskaya, Yu. A. Gorbatkina, A. V. Zamotova, R. L. Kiseleva, T. V. Odnoletkova, and R. Ya. Khvitivitskii, *Mekhanizm Polimerov, Akad. Nauk Latv. SSR,* **1965**, pp 93-99; *Chem. Abstr.,* **63**, 1938h (1965); 1, 4, 14, 22, 24, 25.

(9) G. D. Andreevskaya, Yu. A. Gorbatkina, N. B. Guseva, B. A. Kiselev, A. I. Mikhal'skii, and V. N. Stepanova, *Vysokomolekul. Soedin.,* **7**(7), 1254-7 (1965); *Chem. Abstr.,* **63**, 10123b (1965); 5, 14, 24.

(10) Anon., *Plastics Technol.,* **9**, No. 2, 11-13 (1963); *Chem. Abstr.,* **58**, 9287b (1963); 7, 12, 16, 17, 25.

*Not all references given here are cited in the text, but the subject matter of each reference is indicated by the code numbers following each reference.

(11) C. G. Armistead, and J. A. Hockey, *Trans. Faraday Soc.,* **63**(10), 2549-56 (1967); *Chem. Abstr.,* **67**, 103088x (1967); 14, 24, 32.
(12) R. R. Annand, R. M. Hurd, and N. Hackerman, *J. Electrochem. Soc.,* **112**(2), 138-44 (1965); *Chem. Abstr.,* **62**, 6208h (1965); 13, 23, 32.
(13) M. S. Aslanova, *Plasticheskie Massy,* (8), 31-3 (1960); *Chem. Abstr.,* **55**, 13797c (1961); 7, 14, 16, 25, 32, 33.
(14) M. S. Aslanova, *Verres Refractaires,* **19**(2), 107-12 (1965); *Chem. Abstr.,* **63**, 6693b (1965); 14, 32.
(15) E. C. Atwell, U.S. Pat. 3,318,757 (May 1967); *Chem. Abstr.,* **67**, 22561u (1967); 7, 14, 16, 24.
(16) BTR industries Ltd., French Pat. 1,356,819 (Mar. 1964); *Chem. Abstr.,* **62**, 4220a (1965); 14, 21, 22, 24, 25.
(17) V. F. Ya. Babich, D. Avrasin, and A. L. Rabinovich, *Fiz.-Khim. Mekh. Orientirovannykh Stekloplast,* **1965**, 144-9; *Chem. Abstr.,* **68**, 78975v (1968); 1, 24, 32.
(18) R. E. Baiser, E. G. Shafrin, and W. A. Zisman, *Science,* **162**(3860), 1360-8 (1968); *Chem. Abstr.,* **70**, 48284t (1969); 32, 33.
(19) A. Baigozhin and L. V. Sergeev, *Vysokomolekul. Soedin.,* **4**, 972-6 (1962); *Chem. Abstr.,* **59**, 807g (1963); 4, 14, 24.
(20) A. Baigozhin, L. V. Sergeev, A. K. Dabagova, and S. G. Fattakhov., *Vysokomolekul. Soedin.,* **6**, 75-8 (1963); *Chem. Abstr.,* **60**, 9422h (1964); 7, 14, 24, 32.
(21) G. J. Barenholtz, I. C. Chu, and John E. Rutzler, Jr., *Am. Chem. Soc. Div. Paint, Plastics, Printing Ink, Chem., Preprints,* **20**,(1), 117-23 (1960); *Chem. Abstr.,* **57**, 6120d (1962); 1, 12, 32, 34.
(22) G. J. Barenholtz, J. E. Rutzler, Jr., and E. G. Bobalek, *Ind. Eng. Chem., Prod. Res. Develop.,* **2**(4) 270-2 (1963); *Chem. Abstr.,* **59**, 15436g (1963); 1, 13, 31, 32.
(23) G. J. Barenholtz and J. E. Rutzler, Jr., *Am. Chem. Soc., Div. Org. Coatings Plastics Chem., Preprints,* **22**(1), 223-39 (1962); *Chem. Abstr.,* **59**, 12982c (1963); 1, 13, 31, 32.
(24) F. R. Barnet and J. Cuevas, *Mod. Plastics,* **44**(4), 161-205 (1966); *Chem. Abstr.,* **66**, 38373z (1967); 1, 14, 16, 24, 34.
(25) R. Barnet and J. Cuevas, *Proc. Ann. Tech. Conf.,* SPI *Reinforced Plastics Div. 21st, Chicago,* 1966; *Chem. Abstr.,* **65**, 15600 (1966); 1, 14, 17, 32, 34.
(26) A. J. Barry, U.S. Pat. 2,557,778 (June 1969); *Chem. Abstr.,* **45**, 8288g (1951); 3, 6, 14, 15, 24.
(27) R. Barthel, H. Gocke, and C. M. Hiller, *Plaste Kautschuk,* **14**(12), 894-8 (1967); *Chem. Abstr.,* **68**, 78990w (1968); 4, 5, 7, 14, 15, 24.

(28) E. E. Barton, *Plastics Inst. Trans.*, **18**, 1-21 (1950); *Chem. Abstr.*, **45**, 4479d (1951); 5, 13, 34, 35.
(29) W. D. Bascom, L. A. Halper, and N. L. Jarvis, *Ind. Eng. Chem. Prod. Res. Develop.*, **8**(2), 118-123 (1969); 13, 16, 24, 25, 32.
(30) R. Becker, A. Gesierich, and A. Wende, German Pat. 58,348 (Oct. 1967); *Chem. Abstr.*, **69**, 52717h (1968); 1, 13, 17, 24, 31.
(31) R. Becker, A. Gesierich, A. Wende, and G. Walter, German Pat. 54,049 (Feb. 1967); *Chem. Abstr.*, **68**, 69866h (1968); 1, 14, 24, 34.
(32) R. Becker, A. Gesierich, G. Walter, and A. Wende, German Pat. 41,469 (Sept. 1965); *Chem. Abstr.*, **64**, 8419f (1966); 1, 14, 17, 24.
(33) R. Becker, A. Gesierich, A. Walter, and A. Wende, *Plaste Kautschuk*, **14**(2), 96-8 (1967); *Chem. Abstr.*, **66**, 66244b (1967); 1, 14, 24.
(34) R. Becker, A. Gesierich, I. Schweitzer, and A. Wende, *Plaste Kautschuk*, **14**(6), 406-8 (1967); *Chem. Abstr.*, **67**, 44467v (1967); 1, 6, 14, 24, 34.
(35) R. Becker, *Plaste Kautschuk*, **14**(11), 833 (1967); *Chem. Abstr.*, **68**, 69625 (1968); 1, 16, 17, 24.
(36) R. Becker, A. Gesierich, and A. Wende, *Silikattechnik*, **19**(6), 184-7 (1968); *Chem. Abstr.*, **69**, 79819e (1968); 1, 14, 24.
(37) P. Blatz, U.S. Pat. 3,057,753 (Oct. 1962); *Chem. Abstr.*, **58**, 3576e (1963); 4, 16, 25.
(38) E. Bock, *Angew. Makromol. Chem.*, (4/5), 116-41 (1968); *Chem. Abstr.*, **70**, 205469f (1969); 2, 3, 15, 32.
(39) W. Bollinger and W. Langer, German Pat. 1,285,694 (Dec. 1968); *Chem. Abstr.*, **70**, 69353e (1969); 7, 14, 25.
(40) W. M. Boyer, H. M. Molotsky, and H. E. Tarbell, Jr., U.S. Pat. 3,038,875 (June 1962); *Chem. Abstr.*, **57**, 7413c (1962); 1, 6, 24.
(41) D. M. Brewis, *Polymer Eng. Sci.*, **7**(1), 17-20 (1967); *Chem. Abstr.*, **66**, 46715c (1967); 1, 7, 17, 32.
(42) H. C. Brill, U.S. Pat. 2,943,955 (July 1960); *Chem. Abstr.*, **55**, 2202h (1961); 12, 13, 14, 16, 25.
(43) H. C. Brill, U.S. Pat. 3,002,854 (Appl. Apr. 1957); *Chem. Abstr.*, **56**, 1622i (1962); 4, 12, 16, 25, 31.
(44) Z. I. Bronshtein, Yu. V. Meitin, F. S. Smel'nitskii, A. M. Voronova, and V. A. Murav'ev, *Plasticheskie Massy*, (10), 59-63 (1965); *Chem. Abstr.*, **63**, 18361c (1965); 5, 14, 22, 24.
(45) M. C. Brooks, U.S. Pat. 2,754,237 (July 1956); *Chem. Abstr.*, **50**, 16183f (1956); 4, 14, 24.
(46) M. C. Brooks, U.S. Pat. 2,885,383 (May 1959); *Chem. Abstr.*, **53**, 15651d (1959); 4, 14, 24, 31.

(47) M. C. Brooks, U.S. Pat. 2,891,885 (June 1959); *Chem. Abstr.*, **53**, 19453t (1959); 4, 14, 24.
(48) L. J. Broutman, *Polymer Eng. Sci.*, **6**(3), 263-72 (1966); *Chem. Abstr.*, **65**, 10738g (1966); 1, 4, 14, 25, 33.
(49) A. M. Bueche and C. S. Oliver, U.S. Pat. 3,024,146 (Mar. 1962); *Chem. Abstr.*, **57**, 1098a (1962); 6, 12, 24, 31.
(50) A. M. Bueche and C. S. Oliver, U.S. Pat. 3,031,366 (Apr. 1962); *Chem. Abstr.*, **57**, 4890e (1962); 6, 12, 24, 34.
(51) Burns & Russell Co., British Pat. 966,864 (Aug. 1964); *Chem. Abstr.*, **62**, 1800a (1965); 4, 17, 22, 24.
(52) T. N. Burushkina, A. A. Chuiko, N. V. Khaber, and L. V. Manchenko, *Teor. Eksp. Khim.*, **4**(4), 570-2 (1968); *Chem. Abstr.*, **69**, 99760g (1968); 24, 32.
(53) W. F. Busse and J. A. Boxler, U.S. Pat. 2,838,437 (June 1958); *Chem. Abstr.*, **52**, 17814i (1958); 7, 25.
(54) (a) B. I. Carlstrom and P. T. Agren, Swedish Pat. 198,775 (Oct. 1965); *Chem. Abstr.*, **64**, 9180e (1966); 6, 14, 22, 24. (b) P. E. Cassidy, G. C. Rolls and J. M. Johnson, TRACOR, Inc., TRACOR Document No. T70-AU-7362-U; NASA Contract NAS8-24073. (c) P. E. Cassidy and K. E. Feith, TRACOR Document No. 69-502-C, May 16, 1969, Contract NOOO24-68-C-1227.
(55) R. N. Chadha, French Pat. 1,503,341 (Nov. 1967); *Chem. Abstr.*, **69**, 87881j (1968); 6, 16, 17, 24.
(56) Chi-Chien Ch'eng and Chih-Yao Fang, *Kuei Suan Yen Hsueh Pao*, **4**(3), 161-70 (1965); *Chem. Abstr.*, **64**, 17220c (1966); 14, 24.
(57) P. J. Chevalier, U.S. Pat. 2,635,084 (Apr. 1953); *Chem. Abstr.*, **47**, 8413f (1953); 6, 17, 24.
(58) Uyn-Shih Chi, *Hua Hsueh Tung Pao*, **2**, 97-102 (1962); *Chem. Abstr.*, **57**, 6159a (1962); 7, 14, 24, 32, 33.
(59) A. A. Chuiko, V. A. Tertykh, G. E. Plavnik, and I. E. Neimark, *Kolloidn. Zh.*, **27**(6), 903-7 (1965); *Chem. Abstr.*, **64**, 6825d (1966); 14, 24, 32, 34.
(60) A. A. Chuiko and E. A. Chuiko, *Sintz i Fiz.-Khim. Polimerov, Akad. Nauk Ukr. SSR, Inst. Khim. Vysokomolekul. Soedin., Sb. Statei*, 83-98 (1964); *Chem. Abstr.*, **62**, 9327e (1965); 15, 24, 34.
(61) E. A. Chuiko and A. A. Chuiko, *Sin. Fiz.-Khim. Polimerov, Akad. Nauk Ukr. SSR, Inst. Khim. Vysokomol. Soedin*, **3**, 129-33 (1966); *Chem. Abstr.*, **67**, 100788q (1967); 15, 24, 34.
(62) H. A. Clark and E. P. Plueddemann, *Mod. Plastics*, **40**(10), 133-5, 137-8, 195-6 (1963); *Chem. Abstr.*, **59**, 6579a (1963); 14, 16, 24, 32.

(63) H. A. Clark and E. P. Plueddemann, Belgian Pat. 616,106 (Oct. 1962); *Chem. Abstr.,* **58**, 6971e (1963); 4, 14, 24, 25.
(64) H. A. Clark, U.S. Pat. 2,762,717 (Sept. 1956); *Chem. Abstr.,* **51**, 3955h (1957); 4, 14, 24.
(65) R. E. Clayton and B. M. Vanderbilt, U.S. Pat. 3,376,188 (Apr. 1968); *Chem. Abstr.,* **68**, 96664x (1968); 7, 14, 15, 24.
(66) T. J. Collier, U.S. Pat. 2,974,062 (Mar. 1961); *Chem. Abstr.,* **55**, 20519h (1961); 1, 14, 24.
(67) Continental Gummi-Werke A. G., British Pat. 929,365 (June 1963); *Chem. Abstr.,* **59**, 8946a (1963); 7, 17, 21.
(68) (a) R. P. Cox, L. L. Yaeger, and R. W. Buetow, U.S. Pat. 2,941,912 (June 1960); *Chem. Abstr.,* **54**, 25997i (1960); 7, 16, 21, 34. (b) J. R. Creasey, D. B. Russell, and M. P. Wagner, *Rubber Chem. Technol.,* **41**(5), 1300-11 (1968); *Chem. Abstr.,* **70**, 48399j, (1969); 4, 15, 17, 25.
(69) L. M. Dadson, U.S. Pat. 2,930,715 (Mar. 1960); *Chem. Abstr.,* **54**, 15991h (1960); 15, 22.
(70) C. Dathe, German Pat. 35,226 (Feb. 1965); *Chem, Abstr.,* **63**, 1956g (1965); 6, 12, 24.
(71) C. Dathe, German Pat. 36,060 (May 1965); *Chem. Abstr.,* **63**, 15066c (1965); 6, 17, 24, 31.
(72) C. Dathe and I. Eckhardt, *Plaste Kautschuk,* **10**(7), 404-6 (1963); *Chem. Abstr.,* **61**, 802g (1964); 6, 13, 24.
(73) A. B. Davydov and Z. G. Ivanova, *Plasticheskie Massy,* (1), 46-8 (1965); *Chem. Abstr.,* **62**, 7938h (1965); 16, 24, 34.
(74) W. Dawson, British Pat. 916,865 (Jan. 1963); *Chem. Abstr.,* **58**, 11538c (1963); 14, 16, 22, 24.
(75) R. B. Dean, *Offic. Dig., J. Paint Technol. Eng.* **36**(473), 664-78 (1964); *Chem. Abstr.,* **62**, 2879h (1965); 1, 16, 32.
(76) (a) N. J. DeLollis, Sandia Corporation Research Report, Albuquerque, N. M., 68-270 (May 1968); 1, 4, 5, 6, 7, 14, 24, 31, 32. (b) N. J. DeLollis and O. Montoya, *J. Appl. Polymer Sci.,* **11**, 983-9 (1967); *Chem. Abstr.,* **67**, 44320s (1967); 1, 5, 7, 25, 32.
(77) B. V. Deryagin and V. Smilga, *J. Appl. Phys.,* **38**(12), 4609-16 (1967); *Chem. Abstr.,* **68**, 16400v (1968); 32.
(78) (a) H. Detering, German Pat. 954,456 (Dec. 1956); *Chem. Abstr.,* **53**, 13657d (1959); 1, 6, 14, 17, 24. (b) P. Dickens, K. Gorke, K. H. Schmidt, C. Kaser, and P. König, German Pat. 1,052,295 (Mar. 1959); *Chem. Abstr.,* **55**, 3947d (1961); 7, 17, 22, 33.

(79) Dow Corning Corp. Netherlands Appl. 6,613,777 (Mar. 1967), U.S. Appl. (Sept. 1965); *Chem. Abstr.,* **67**, 91330e (1967); 6, 17, 24.
(80) Dow Corning Corp., Netherlands Appl. 6,600,136 (July 1966), U.S. Appl. (Jan. 1965); *Chem. Abstr.,* **65**, 20320b (1966); 14, 15, 16, 17, 24, 33.
(81) Dow Corning Corp., Netherlands Appl. 6,511,722 (Mar. 1966), U.S. Appl. (Sept. 1964); *Chem. Abstr.,* **65**, 4028a (1966); 6, 16, 24.
(82) Dow Corning Corp., British Pat. 975,122 (Nov. 1964); *Chem. Abstr.,* **62**, 7948b (1965); 1, 4, 7, 14, 24, 31.
(83) Dow Corning Corp., Netherlands Appl. 6,402,923 (Nov. 1964), U.S. Appl. (May 1963); *Chem. Abstr.,* **62**, 10628a (1965); 7, 16, 24, 31.
(84) Dow Corning Corp., British Pat. 1,002,795 (Aug. 1965); *Chem. Abstr.,* **63**, 18387d (1965); 5, 7, 12, 17, 24.
(85) Dow Corning Product Bulletin, "Silane Coupling Agents," Dow Corning Corp., Chem. Products Div., Midland, Mich.; 1, 2, 3, 4, 5, 6, 7, 12, 13, 14, 24, 31.
(86) R. W. Drisko, *Paint Varn. Prod.,* **58**(7), 31-4 (1968); *Chem. Abstr.,* **69**, 52967g (1968); 1, 16, 17, 24, 31, 34.
(87) E. I. duPont de Nemours & Co., British Pat. 1,107,419 (Mar. 1968); *Chem. Abstr.,* **69**, 3439k (1968); 1, 2, 12, 13, 16, 24.
(88) E. I. duPont de Nemours & Co., British Pat. 923,382 (Apr. 1963); *Chem. Abstr.,* **59**, 8948a (1963); 1, 14, 17, 22, 31, 33.
(89) W. J. Eakins, U.S. Pat. 3,276,853 (Oct. 1966); *Chem. Abstr.,* **65**, 18785d (1966); 14, 24.
(90) H. Eder and S. Bloeck, German Pat. 1,204,111 (Oct. 1965); *Chem. Abstr.,* **64**, 8500h (1966); 12, 24, 33.
(91) G. E. Eilerman, U.S. Pat. 3,025,588 (Mar. 1962); *Chem. Abstr.,* **57**, 1117e (1962); 3, 14, 22, 24.
(92) P. W. Erickson, NASA Accession N65-24572, Rept. No. AD 426292, (1964); *Chem. Abstr.,* **67**, 33348f (1967); 7, 14, 32.
(93) P. W. Erickson and I. Silver, U.S. Pat. 2,776,910 (Jan. 1957); *Chem. Abstr.,* **51**, 7767g (1957); 1, 14, 24.
(94) J. M. Fain, E. McDonnell, and R. Blaufox, U.S. Pat. 3,244,541 (Apr. 1966); *Chem. Abstr.,* **64**, 19999f (1966); 14, 24, 33.
(95) F. Fekete, U.S. Pat. 3,321,350 (May 1967); *Chem. Abstr.,* **67**, 33645a (1967); 3, 6, 12, 13, 14, 24.
(96) F. Fekete, U.S. Pat. 3,296,182 (Jan. 1967); *Chem. Abstr.,* **66**, 56391h (1967); 6, 24, 25, 34.

(97) (a) F. Fekete, British Pat. 887,441 (Jan. 1962); *Chem. Abstr.*, **57**, 1098d (1962); 6, 17, 24, 25. (b) N. F. Finley and N. Hackerman, *J. Electrochem. Soc.*, **107**, 259 (1960); 23, 32.
(98) C. E. Frank and G. Kraus, U.S. Pat. 2,692,841 (Oct. 1954); *Chem. Abstr.*, **49**, 2770e (1955); 15, 17, 32, 33.
(99) K. Franzen and G. Herkelmann, German Pat. 1,254,303 (Nov. 1967); *Chem. Abstr.*, **68**, 13786q (1968); 3, 5, 14, 24. 25.
(100) Fujimori Kogyo Co., Ltd., French Pat. 1,449,532 (Aug. 1966), Japan Appl. (Oct. 1964); *Chem. Abstr.*, **66**, 96402u (1967); 7, 17, 22, 25, 33.
(101) J. Gaehde and A. Wende, *Plaste Kautschuk*, **13**(1), 23-5 (1966); *Chem. Abstr.*, **64**, 11387b (1966); 7, 14, 16, 24, 32.
(102) (a) D. Gagliardi, British Pat. 1,020,052 (Feb. 1966); *Chem. Abstr.*, **64**, 16055e (1966); 17, 24, 33. (b) General Aniline and Film Corp., "Gantrez" Product Bulletin, 140 W. 51 St., N.Y., N.Y.; 7, 22.
(103) General Electric Co., British Pat. 1,091,650 (Nov. 1967); *Chem. Abstr.*, **68**, 22572u (1968); 16, 24, 33.
(104) General Electric Co., British Pat. 825,717 (Oct. 1960); *Chem. Abstr.*, **55**, 9877d (1961); 16, 17, 24.
(105) J. J. Giordano, French Pat. 1,439,853 (May 1966); *Chem. Abstr.*, **66**, 66344j (1967); 7, 12, 13, 14, 24.
(106) Th. Goldschmidt, British Pat. 930,039 (July 1963); *Chem. Abstr.*, **59**, 14174g (1963); 14, 24, 32.
(107) L. I. Golubenkova, S. N. Nikonova, A. N. Shabadash, and M. S. Akutin, *Appl. Plastics*, **10**(5), 44-6, 48-9 (1967); *Chem. Abstr.*, **67**, 33262y (1967); 1, 4, 14, 24, 32.
(108) J. T. Goodwin, Jr., U.S. Pat. 2,857,356 (Oct. 1958); *Chem. Abstr.*, **53**, 1857e (1959); 6, 14, 16, 24.
(109) Yu. A. Gorbatkina, *Plaste Kautschuk*, **13**(3), 145-50 (1966); *Chem. Abstr.*, **64**, 17796e (1966); 1, 5, 6, 7, 14, 22, 25, 33.
(110) Yu. A. Gorbatkina, N. B. Guseva, G. D. Andreevskaya, and G. S. Galakhova, *Vysokomolekul. Soedin.*, **6**(10), 1911-16 (1964); *Chem. Abstr.*, **62**, 2881h (1965); 1, 5, 7, 14, 24, 31.
(111) A. Gottfurcht, U.S. Pat. 2,763,629 (Sept. 1956); *Chem. Abstr.*, **51**, 3105d (1957); 7, 14, 24.
(112) C. L. Gray, Jr., W. L. MacCarthy, and T. F. McLaughlin, *Mod. Packaging*, **34**, No. 10, 143-47, 243-5 (1961); *Chem. Abstr.*, **55**, 20494h (1961); 12, 16, 25.
(113) C. J. Gray, Jr., *Am. Chem. Soc., Div. Org. Coatings Plastics Chem., Preprints*, **20**, No. 2, 382-95 (1960); *Chem. Abstr.*, **58**, 11529g (1963); 4, 12, 16, 25.

(114) K. P. Grinevich, N. N. Kryuchkov, M. V. Sobolevskii, Z. I. Bronshtein, S. N. Odisharia, M. N. Krichevskaya, and A. V. Bosomykina, U.S.S.R. Pat. 144,284 (Feb. 1962); *Chem. Abstr.*, **57**, 6126d (1962); 4, 14, 24.

(115) (a) H. R. Guivier and R. M. Joly, French Pat. 1,508,381 (Jan. 1968); *Chem. Abstr.*, **70**, 20641h (1969); 1, 17, 24. (b) N. Hackerman, R. M. Hurd, and R. P. Annand, *Corrosion*, **18**, 37t (1962); 23, 32.

(116) K. F. Hager and M. Rosenthal, U.S. Pat. 2,853,412 (Sept. 1958); *Chem. Abstr.*, **53**, 1855h (1959); 6, 12, 13, 24, 31.

(117) M. L. Hair and W. Hertl, *J. Phy. Chem.*, **73**(7), 2372 (1969); 14, 24, 32.

(118) V. E. Hamilton and L. M. Roseland, S. African Pat. 67 04,865 (Jan. 1968); *Chem. Abstr.*, **70**, 38462k (1969); 1, 2, 14, 24.

(119) J. R. Harper, French Pat. 1,386,548 (Jan. 1965); *Chem. Abstr.*, **63**, 10173f (1965); 6, 14, 16, 17, 24.

(120) J. R. Harper, U.S. Pat. 2,979,420 (Apr. 1961); *Chem. Abstr.*, **55**, 21640e (1961); 6, 12, 13, 14, 24.

(121) F. B. Hauserman, *Advan. Chem. Ser.*, **23**, 338-56 (1959); *Chem. Abstr.*, **54**, 4234a (1960); 22, 29.

(122) L. Havestadt and H. Arens, German Pat. 800,685 (Nov. 1950); *Chem. Abstr.*, **45**, 3142e (1951); 7, 17, 24.

(123) J. J. Hazdra and H. M. Lontz, U.S. Pat. 3,341,399 (Sept. 1967); *Chem. Abstr.*, **67**, 100720m (1967), 7, 14, 24, 31.

(124) R. M. Hedrick, W. R. Richard, Jr., A. J. Lambright, and R. E. Miller, Belgian Pat. 648,690 (Dec. 1964); *Chem. Abstr.*, **63**, 13504d (1965); 6, 16, 24.

(125) Henkel & Cie, G.m.b.H., British Pat. 989-390 (Apr. 1965); *Chem. Abstr.*, **63**, 1960e (1965); 1, 12, 15, 25.

(126) A. R. Henning, *Kunststoffe* **44**, 131-3 (1954); *Chem. Abstr.*, **48**, 7925h (1954); 4, 14, 22, 24.

(127) J. Hetflejs, F. Mares, and V. Chvalovsky, *Intern. Symp. Organosilicon Chem., Sci. Commun., Prague, 1965,* 282-9; *Chem. Abstr.*, **66**, 2122c (1967); 24, 32.

(128) C. M. Hiller, O. Quast, and G. Wiedemann, *Plaste Kautschuk,* **11**, 136-40 (1964); *Chem. Abstr.*, **63**, 10111a (1965); 5, 14, 22, 24.

(129) W. Hinz and G. Solow, *Silikat Tech.*, **8**, 178-85 (1957); *Chem. Abstr.*, **51**, 17225a (1957); 4, 14, 22, 24, 32, 35.

(130) P. Hittmair, E. Wohlfarth, S. Nitsche, and M. Wick, U.S. Pat. 3,432,463 (Mar. 1969); *Chem. Abstr.*, **70**, 97804w (1969); 6, 24.

(131) F. Hoersch and H. Wurtinger, *Kunststoffe* **56**(9), 627-9 (1966); *Chem. Abstr.*, **65**, 20284e (1966); 1, 14, 22, 24, 31.

(132) R. Holtmann, *Kunststoffe*, **57**(2), 110-16 (1967); *Chem. Abstr.*, **67**, 33191z; 1, 4, 14, 24, 31.
(133) Hooker Chemical Corp., Netherlands Appl. 6,400,517 (July 1964), U.S. Appl. (Jan. 1963); *Chem. Abstr.*, **62**, 4174b; 1, 21.
(134) Howmet Corp., British Pat. 1,101,000 (Jan. 1968); *Chem. Abstr.*, **68**, 89487p (1968); 24, 33.
(135) J. R. Huntsberger, *Advan. Chem. Ser.*, **43**, 180-8 (1964); *Chem. Abstr.*, **60**, 10811b (1964); 31.
(136) J. R. Huntsberger, *J. Paint Technol.*, **39**(507), 199-211 (1967); *Chem. Abstr.*, **67**, 3358s (1967); 32.
(137) B. Hurley, E. L. Smith, W. Kitaj, and R. H. Kiel, Belgian Pat. 612,052 (June 1962); *Chem. Abstr.*, **57**, 12122a (1962); 7, 14, 25.
(138) W. E. Hutchinson, U.S. Pat. 3,024,131 (Mar. 1962); *Chem. Abstr.*, **56**, 14430h (1962); 7, 15, 17, 24.
(139) J. Iannicelli, U.S. Pat. 3,290,165 (Dec. 1966); *Chem. Abstr.*, **66**, 105491b (1967); 16, 24, 33, 34.
(140) Imperial Chemical Ind. Ltd., Netherlands Appl. 6,516,715 (Apr. 1966), British Appl. (Dec. 1964, June 1965, Dec. 1965); *Chem. Abstr.*, **65**, 10751e (1966); 1, 14, 16, 24.
(141) A. Isirikyan, A. V. Kiselev, and E. V. Ushakova, *Kolloidn. Zh.*, **27**(5), 690-6 (1965); *Chem. Abstr.*, **64**, 3841c (1966); 17, 24, 32.
(142) Z. K. Jelinek, Czech. Pat. 89,064 (Mar. 1959); *Chem. Abstr.*, **55**, 23967e (1961); 14, 22.
(143) E. Jenckel and H. Huhn, *Kolloid-Z.*, **159**, 118-29 (1958); *Chem. Abstr.*, **53**, 28a (1959); 7, 12, 25.
(144) M. Joesler, *Atti Congr. Intern. Materie Plastiche*, **14**, 228-36 (1962); *Chem. Abstr.*, **59**, 14162f (1963); 1, 4, 5, 7, 14, 16, 24, 32.
(145) O. K. Johannson, F. O. Stark, G. E. Vogel, and R. M. Fleischmann, *J. Compos. Mater.*, **1**(3), 278-92 (1967); *Chem. Abstr.*, **67**, 82642f (1967); 7, 14, 24, 32.
(146) O. K. Johannson, U.S. Pat. 2,557,786 (June 1951); *Chem. Abstr.*, **45**, 8222g (1951); 7, 14, 24, 31.
(147) A. C. Johnson and R. C. Shaffer, U.S. Pat. 3,407,169 (Oct. 1968); *Chem. Abstr.*, **70**, 20773c (1969); 1, 5, 6, 24, 34.
(148) H. L. Jones, A. N. Cianciarulo, and R. L. Webb, *Am. Chem. Soc., Div. Org. Coatings, Plastics Chem., Preprints*, **24**(2), 351-9 (1964); *Chem. Abstr.*, **64**, 12877b (1966); 1, 14, 32.
(149) L. C. Jones, U.S. Pat. 3,117,900 (Jan. 1964); *Chem. Abstr.*, **60**, 12235a (1964); 1, 4, 17, 24.

(150) K. Kadowaki and T. Nakagiri, Japanese Pat. 68 13,996 (June, 1968): *Chem. Abstr.*, **69**, 107351s (1968); 7, 16, 25.
(151) E. I. Kalinina and E. M. Lukina, U.S.S.R. Pat. 173,705 (Aug. 1965); *Chem. Abstr.*, **64**, 2249g (1966); 7, 22.
(152) (a) G. Kaliske, *Dissertationen,* No. 21, sheet 64.14 (1964); *Chem. Abstr.*, **64**, 2233d (1966); 5, 31. (b) M. Kalnins and V. Karlivane, *Latv. PSR Zinat. Akad. Vestis, Kim. Ser.*, **3**, 292-5 (1968); *Chem. Abstr.*, **69**, 78040g (1968); 7, 13, 16, 25.
(153) (b) B, Kanner, U.S. Pat. 3,247,280 (Apr. 1966); *Chem. Abstr.*, **64**, 19906c (1966); 1, 6, 24.
(154) B. Kanner, British Pat. 906,544 (Sept. 1962); *Chem. Abstr.*, **58**, 3575g (1963); 1, 6, 24.
(155) C. J. Kastrop, U.S. Pat. 3,160,675 (Dec. 1964); *Chem. Abstr.*, **62**, 6641f (1965); 1, 5, 16, 17, 24, 31.
(156) J. W. Keil, U.S. Pat. 2,902,389 (Sept. 1959); *Chem. Abstr.*, **54**, 1920g (1960); 6, 12, 17, 24.
(157) A. S. Kenyon, *J. Colloid Interface Sci.*, **27**(4), 761-71 (1968); *Chem. Abstr.*, **69**, 97398 (1968); 1, 14, 32.
(158) B. A. Kisley, A. I. Mikhal'skii, V. V. Bodrova, and N. G. Korbut, *Fiz.-Khim. Mekh. Orientirovannykh Stekloplast., Sb. Rab. Konf., Moscow,* 30-43 (1965); *Chem. Abstr.*, **68**, 78997d (1968); 7, 14, 24, 32.
(159) F. F. Koblitz, U.S. Pat. 2,774,697 (Dec. 1956); *Chem. Abstr.*, **51**, 4055g (1957); 7, 12, 14, 16, 24.
(160) K. E. Kolb and J. G. Koelling, *Proc. Ann. Tech. Congr. SPI Reinforced Plastics Div., 21st, Chicago, 1966*; *Chem. Abstr.*, **67**, 12045a (1967); 1, 14, 24, 31, 32.
(161) S. Kori, *Kyoka Purasuchikkusu,* **6**, 54-60 (1960); *Chem. Abstr.*, **57**, 16153a (1962); 14, 22, 24, 32.
(162) S. Koristek, J. Hajek, and J. Stepanek, Czech. Pat. 120,411 (Nov. 1966); *Chem. Abstr.*, **67**, 102503s (1967); 4, 14, 24.
(163) A. Ya. Korolev, P. V. Davydov, and L. N. Vinogradova, *Vysokomolekul. Soedin., Adgeziya Polimerov, Sb. Statei,* 3-11 (1963); *Chem. Abstr.*, **60**, 9331e (1964); 2, 13, 14, 24, 32.
(164) M. P. Kotov and N. S. Sorokina, *Izv. Vysshikh Uchebn. Zavedeniic, Tekhnol. Legkoi Prom.*, **1968**(6), 66-9; *Chem. Abstr.*, **70**, 58674j (1969); 32.
(165) L. M. Koval'chuk and A. M. Chistyokov, *Plasticheskie Massy,* (4), 65-6 (1968); *Chem. Abstr.*, **69**, 3529a (1968); 1, 12, 31.

(166) M. Kramer, W. Noll, and O. Pantke, German Pat. 1,080,061 (Apr. 1960); *Chem. Abstr.,* **55**, 18132d (1961); 17, 22, 24.
(167) N. A. Krotova, L. P. Morozova, and B. V. Deryagin, *Doklady Akad. Nauk SSSR,* **129**, 149-52 (1959); *Chem. Abstr.,* **54**, 8220a (1960); 7, 14, 24, 32.
(168) N. A. Krotova and L. P. Morozova, *Issled. v Obl. Poverkhn. Sil, Akad. Nauk SSSR, Inst. Fiz. Khim., Sb. Dokl. na Vtoroi Konf., Moscow,* 322-9 (1962); *Chem. Abstr.,* **62**, 4129b (1965); 7, 14, 24, 25, 32.
(169) N. A. Krotova, L. P. Morozova, A. M. Polyakov, G. A. Sokolina, and N. N. Stefanovich, *Kolloidn. Zh.,* **26**(2), 207-14 (1964); *Chem. Abstr.,* **61**, 7181h (1964); 32.
(170) N. A. Krotova, L. P. Morozova, and A. M. Polyakov, *Probl. Fiz-Khim. Mekh. Voloknistykh Poristykh Dirpersnykh Strukt. Mater., Mater. Konf., Riga,* 497-507 (1965); *Chem. Abstr.,* **68**, 115264k (1968); 14, 15, 32.
(171) B. K. Krys'kova, A. M. Simonenko, L. N. Ol'shevskaya, and M. V. Varentsova, *Khim. Obrab. Poverkh. Steklyannogo Volokna, Vses. Nauch.-Issled. Inst. Stekloplastikov Steklyannogo Volokna,* 21-6 (1968); *Chem. Abstr.,* **68**, 88088x (1968); 14, 24, 32.
(172) B. P. Kryzhanovskii, A. Ya. Kuznetsov, G. V. Okatova, and M. A. Okatov, *Opt.-Mekh. Prom.,* **35**(2), 45-8 (1968); *Chem. Abstr.,* **69**, 28429v (1968); 3, 6, 14, 24.
(173) V. A. Kudishina, *Vestn. Mashinostr,* **48**(7), 44-5 (1968); *Chem. Abstr.,* **70**, 20714j (1969); 7, 13, 14, 17, 24, 25.
(174) C. A. Kumins, W. M. Arnheim, and T. K. Kwei, *Proc. Ann. Tech. Management Conf. Reinforced Plastics Div., Soc. Plastics Ind.,* **19**, (11-D) (1964); *Chem. Abstr.,* **61**, 16249a (1964); 14, 22, 31, 32.
(175) V. P. Kuznetsova, S. I. Omel'chenko, G. N. Belogolovina, and V. G. Smelova, *Sin. Fiz.-Khim. Polim.,* No. 5, 162-7 (1968); *Chem. Abstr.,* **70**, 5258v (1969); 2, 24.
(176) J. A. Laird and F. W. Nelson, *Proc. Ann. Tech. Management Conf., Reinforced Plastics Div., Soc. Plastics Ind.,* **19**(11-C), (1964); *Chem. Abstr.,* **61**, 16251c (1964); 1, 14, 24, 31.
(177) R. D. Lawrence and D. H. Geschwind, U.S. Pat. 3,337,485 (Aug. 1967); *Chem. Abstr.,* **68**, 79066m (1968); 7, 12, 13, 17, 25.
(178) L-H., Lee, *Am. Chem. Soc., Div. Polymer Chem., Preprints,* **7**(2), 910-15 (1966); *Chem. Abstr.,* **66**, 76732y (1967); 3, 14, 17, 32.
(179) A. F. Lewis and L. J. Forrestal, *Am. Soc. Testing Mater., Spec. Tech. Publ. 360,* **(1963)** 59-75; *Chem. Abstr.,* **62**, 5389b (1965); 7, 11, 12, 13, 21, 25, 32.

(180) Yu. S. Lipatov, *Modifkatsiya Svoistv Polimerov i Polimernykh Materialov, Akad. Nauk Ukr. SSR, Resp. Mezhvedomstv. Sb.*, **1965**, 56-63; *Chem. Abstr.*, **65**, 2418h (1966); 7, 14, 16, 34.
(181) Yu. S. Lipatov, *Plast. Inst., Trans. J.*, **34**(110), 83-7 (1966); *Chem. Abstr.*, **65**, 10734f (1966); 14, 16, 32, 33.
(182) Lord Manufacturing Co., British Pat. 850,419 (Oct. 1960); *Chem. Abstr.*, **55**, 9946e (1961); 6, 12, 13, 17, 24.
(183) V. B. Losev, K. B. Ryazonov, and V. D. Orlova, *Plasticheskie Massy*, (2), 46 (1967); *Chem. Abstr.*, **66**, 86179n (1967); 1, 6, 17, 24.
(184) A. H. Lund, U.S. Pat. 3,177,170 (Apr. 1965); *Chem. Abstr.*, **63**, 801e (1965); 1, 4, 7, 14, 24.
(185) V. P. Lyutyi, N. P. Kharitonov, and L. P. Shtremt, *Zh. Prikl. Khim.*, **38**(5), 1131-3 (1965); *Chem. Abstr.*, **63**, 5829g (1965); 1, 14, 24, 34.
(186) G. P. Maly, U.S. Pat. 3,146,142 (Aug. 1964); *Chem. Abstr.*, **61**, 12163b (1964); 1, 5, 7, 14, 17, 24, 25.
(187) H. Manami and S. Nishizaki, *Kogyo Kagaku Zasshi*, **61**, 218-19 (1958); *Chem. Abstr.*, **54**, 6171e (1960); 6, 17, 24, 34.
(188) T. J. Mao and S. L. Reegan, *Proc. Symp. Adhesion Cohesion, 5th, Warren, Mich., 1961*, 209-17; *Chem. Abstr.*, **59**, 14121g (1963); 7, 17, 32.
(189) Yu. I. Markin, V. M. Gorchakova, V. E. Gul, and S. S. Voyutskii, *Izv. Vysshikh Uchebn. Zavedenii, Khim. i Khim. Tekhnol.*, **5**, 808-14 (1962); *Chem. Abstr.*, **58**, 12693f (1963); 7, 11, 12, 17, 32.
(190) J. G. Marsden and S. Sterman, French Pat. 1,462,596 (Dec. 1966); *Chem. Abstr.*, **67**, 93658s (1967); 14, 22, 24, 33.
(191) A. Marzocchi, U.S. Pat. 3,287,204 (Nov. 1966); *Chem. Abstr.*, **66**, 19594f (1967); 3, 14, 15, 24.
(192) A. Marzocchi and N. S. Janetos, U.S. Pat. 3,252,825 (May 1966); *Chem. Abstr.*, **65**, 7465c (1966); 7, 14, 24.
(193) A. Marzocchi and N. S. Janetos, U.S. Pat. 3,169,884 (Feb. 1965); *Chem. Abstr.*, **62**, 14307e (1965); 1, 14, 24.
(194) A. Marzocchi and N. S. Janetos, U.S. Pat. 3,207,623 (Sept. 1965); *Chem. Abstr.*, **63**, 18344a (1965); 4, 14, 24.
(195) A. Marzocchi and N. S. Janetos, German Pat. 1,141,788 (Dec. 1962); *Chem. Abstr.*, **58**, 14225f (1963); 1, 14, 24, 25.
(196) A. Marzocchi and N. S. Janetos, U.S. Pat. 2,931,739 (Apr. 1960); *Chem. Abstr.*, **54**, 16921g (1960); 1, 4, 14, 24.
(197) J. E. Masters, U.S. Pat. 3,150,116 (Sept. 1964); *Chem. Abstr.*, **61**, 16264g (1964); 1, 24.
(198) V. A. Matonis, *Polymer Eng. Sci.*, **9**(2), 100 (1969); 24, 32.

(199) A. Matting and W. Brockman, *Adhaesion,* (8), 343-51 (1968); *Chem. Abstr.,* **70**, 78619y (1969); 5, 32, 33.
(200) W. F. McWhorter, U.S. Pat. 3,154,597 (Oct. 1964); *Chem. Abstr.,* **62**, 1811a (1965); 1, 14, 24.
(201) Yu. V. Meitin, *Plasticheskie Massy,* (12), 48-53 (1964); *Chem. Abstr.,* **62**, 6610d (1965); 14, 22, 24, 32.
(202) Michelin, During & Cie, French Pat. 1,278,030 (Mar. 1962); *Chem. Abstr.,* **57**, 4890h (1962); 1, 16, 17, 25.
(203) Midland Silicones Ltd., British Pat. 813,889 (May 1959); *Chem. Abstr.,* **53**, 19464f (1959); 6, 14, 24.
(204) Minnesota Mining and Mfg. Co., Belgian Pat. 654,206 (Feb. 1965); *Chem. Abstr.,* **64**, 14418f (1966); 1, 14, 24.
(205) W. Moebes and A. Wende, *Plaste Kautschuk,* **9**, 232-6 (1962); *Chem. Abstr.,* **63**, 4468c (1965); 6, 14, 24, 32.
(206) Monsanto Co., Netherlands Appl. 6,415,099 (June 1965); *Chem. Abstr.,* **64**, 892a (1966); 4, 7, 17, 21, 34.
(207) Monsanto Co., Netherlands Appl. 6,413,905 (Dec. 1965); *Chem. Abstr.,* **64**, 16016a (1966) 24, 34.
(208) Monsanto Co., Netherlands Appl. 6,600,164 (July 1966); *Chem. Abstr.,* **66**, 19216j (1967); 24, 34.
(209) A. W. Morgan and J. S. Swenson, U.S. Pat. 3,376,278 (April 1968); *Chem. Abstr.,* **68**, 105746e (1968); 7, 16, 25, 33.
(210) W. L. Morgan, U.S. Pat. 3,013,915 (Dec. 1961); *Chem. Abstr.,* **56**, 11807a (1962); 14, 16, 24.
(211) L. P. Morozova and N. A. Krotova, *Issled. Obl. Poverkh. Sil, Sb. Dokl. Konf., 3rd, 1966*, 442-7; *Chem. Abstr.,* **70**, 4852 (1969); 14, 24, 32.
(212) S. N. Muchnick, *Mech. Eng.,* **78**, 19-22 (1956); *Chem. Abstr.,* **50**, 2389f (1956); 12, 17, 32, 33.
(213) U. Mueller and G. Wiedemann, *Faserforsch. Textiltech.,* **20**(1), 10-20 (1969); *Chem. Abstr.,* **70**, 97855p (1969); 14, 22, 24, 32.
(214) H. R. Nack, U.S. Pat. 2,791,515 (May 1957); *Chem. Abstr.,* **51**, 12460c (1957); 17, 22.
(215) T. Nakabayashi, A. Ikari, and J. Tsukamoto, *Chem. High Polymers,* **11**, 123-6 (1954); *Chem. Abstr.,* **50**, 1362g (1956); 5, 25.
(216) R. J. Nebesar, U.S. Pat. 2,607,708 (Aug. 1952); *Chem. Abstr.,* **46**, 11616d (1952); 6, 14, 22.
(217) L. E. Nielson, *J. Appl. Polymer Sci.,* **10**(1), 97-103 (1966); *Chem. Abstr.,* **64**, 11381d (1966); 32, 34.

(218) P. Nitrogenmuvck, U. Kutato Intezet, and N. H. V. Kutato Intezet, Hungarian Pat. 151,460 (June 1964); *Chem. Abstr.*, **61**, 7198c (1964); 7
(219) S. Nitzsche, *Kunststoffe,* **41**, 87-8 (1951); *Chem. Abstr.*, **45**, 5968b (1951); 14, 24.
(220) S. Nitzsche, M. Wick, and E. Wohlfahrth, German Pat. 1,205,279 (Nov. 1965); *Chem. Abstr.*, **64**, 5276g (1966); 6, 14, 17, 24.
(221) S. Nitzsche and M. Wick, U.S. Pat. 3,108,898 (Oct. 1963); *Chem. Abstr.*, **60**, 5722b (1964); 6, 15, 16, 17, 24.
(222) P. Nowak and F. J. Bollig, German Pat. 1,056,979 (May 1959); *Chem. Abstr.*, **55**, 15983g (1961); 6, 15, 17, 24.
(223) S. Orchon, *Tappi,* **40**, No. 9, 182A-184A (1957); *Chem. Abstr.*, **52**, 5879b (1958); 32.
(224) S. Orchon, *Tappi,* **41**, 33-7 (1958); *Chem. Abstr.*, **52**, 10688c (1958); 5, 7, 32.
(225) J. G. O'Rear, P. J. Sniegoski, F. L. James, *Advan. Chem. Ser.*, **1968**, (87), 10-19; *Chem. Abstr.*, **70**, 39108m (1969); 7, 17, 24, 25, 32.
(226) H. P. Owen, *Mater. Symp., 7th, Los Angeles, Calif., No. 10, 1964*; *Chem. Abstr.*, **62**, 5395c (1965); 1, 5, 6, 12, 13, 17, 31.
(227) Owens-Corning Fiberglass Corp., Netherlands Appl. Pat. 6,505,006 (Oct. 1965); *Chem. Abstr.*, **64**, 9954d (1966); 1, 14, 22, 25.
(228) R. E. Park and C. B. Sias, U.S. Pat. 3,334,008 (Aug. 1967); *Chem. Abstr.*, **67**, 74133a (1967); 4, 14, 24.
(229) R. L. Patrick and W. S. Layne, *Proc. Ann. Tech. Conf. SPI Reinforced Plastics Div., 21st, Chicago, 1966, Sec. 13A*; *Chem. Abstr.*, **65**, 15583g (1966); 1, 14, 31, 32.
(230) N. N. Pavlov and G. A. Arbuzov, *Nauchn. Tr. Mosk. Tekhnol. Inst. Legkoi Prom.*, **1960**(17), 29-34; *Chem. Abstr.*, **56** 15672f (1962); 7, 22.
(231) F. S. Perkerson, U.S. Pat. 2,499,774 (Mar. 1950); *Chem. Abstr.*, **44**, 5141a (1950); 3, 15, 17, 25.
(232) C. M. Peterson, *Kolloid-Z. Z. Polymer,* **222**(2), 148-57 (1968); *Chem. Abstr.*, **68**, 115469f (1968); 1, 13, 32.
(233) R. M. Pike, U.S. Pat. 3,018,299 (Jan. 1962); *Chem. Abstr.*, **57**, 13803g (1962); 7, 13, 14, 15, 17, 24, 32.
(234) (a) A. N. Pines, U.S. Pat. 3,083,847 (May 1963); *Chem. Abstr.*, **59**, 4128e (1963); 6, 7, 11, 12, 13, 17, 24. (b) A. N. Pines, German Pat. 1,238,598 (Apr. 1967); *Chem. Abstr.*, **67**, 44967h (1967); 11, 17, 24, 31.

(c) A. N. Pines and S. Sterman, British Pat. 952,992 (Mar. 1964); *Chem. Abstr.*, **61**, 16312h (1964); 1, 6, 17, 24.

(235) E. P. Plueddemann, *J. Paint Technol.*, **40**(516), 1-9 (1968); *Chem. Abstr.*, **68**, 60553w (1968); 14, 17, 24, 29, 32.

(236) E. P. Plueddemann, *Proc. Ann. Tech. Conf. SPI Reinforced Plastics Div., 20th, Chicago, 1965,* Sec. 19A; *Chem. Abstr.*, **63**, 3124c (1965); 7, 14, 16, 24, 29.

(237) E. P. Plueddemann, *Proc. Ann. Tech. Conf. SPI Reinforced Plastics Div., 21st, Chicago, 1966,* Sec. 3D; *Chem. Abstr.*, **65**, 10732a (1966); 7, 14, 24, 31.

(238) E. P. Plueddemann, *Proc. Ann. Tech. Conf. SPI Reinforced Plastics Div. 22nd,* Washington, *1967,* Sec. 9A; *Chem. Abstr.*, **67**, 54674z (1967); 14, 16, 24, 31, 32.

(239) E. P. Plueddemann, French Pat. 1,460,691 (Dec. 1966); *Chem. Abstr.*, **67**, 22542p (1967); 1, 5, 7, 14, 16, 24.

(240) E. P. Plueddemann, French Pat. 1,492,467 (Aug. 1967); *Chem. Abstr.*, **68**, 105796w (1968); 7, 14, 17, 24, 31.

(241) E. P. Plueddemann, H. A. Clark, L. E. Nelson, and K. R. Hoffman, *Mod. Plastics,* **39**, 135 (1962); 1, 4, 14, 24, 31.

(242) G. Pohl, *Plaste Kautschuk,* **15**(8), 565-8 (1968); *Chem. Abstr.*, **69**, 52697b (1968); 4, 14, 24, 32.

(243) K. E. Polmanteer, U.S. Pat. 3,155,633 (Nov. 1964); *Chem. Abstr.*, **62**, 4178h (1965); 6, 13, 15, 24, 34.

(244) M. C. Polniaszek and R. H. Schaufelberger, *Adhesives Age,* **11**(7), 25-7 (1968); *Chem. Abstr.*, **69**, 52694y (1968); 1, 5, 6, 7, 12, 24.

(245) R. E. Powell, British Pat. 1,132,573 (Nov. 1968); *Chem. Abstr.*, **70**, 20871h (1969); 4, 17, 25, 35.

(246) L. M. Prutkov, K. A. Andrianov, N. A. Polikanin, and E. Z. Asnovich, U.S.S.R. Pat. 171,577 (May 1965); *Chem. Abstr.*, **63**, 15067g (1965); 1, 6, 24.

(247) I. H. Riley and H. W. Winnam, British Pat. 1,014, 272 (Dec. 1965); *Chem. Abstr.*, **64**, 8455b (1966); 6, 17, 24.

(248) R. C. Rittenhouse, German Pat. 1,107,860 (May 1960); *Chem. Abstr.*, **56**, 2550f (1962); 1, 15, 17, 24.

(249) R. C. Rittenhouse, British Pat. 952,993, (Mar. 1964); *Chem. Abstr.*, **61**, 5888d (1964); 6, 12, 15, 16, 17, 24.

(250) N. L. Rogers, *Appl. Poly. Symp.*, **3**(1-25), 327-40 (1966); *Chem. Abstr.*, **66**, 88091b (1967), 11, 12, 13, 17.

(251) R. M. Roseland, *Proc. Ann. Tech. Conf., SPI Reinforced Plastics Div., 21st, Chicago, 1966,* Sec. 4C; *Chem. Abstr.,* **65**, 11789e (1966); 24, 31.
(252) Ruetgerswerke and Teerverwertung A.-G., Netherlands Appl. Pat. 6,510,320 (Feb. 1966); *Chem. Abstr.,* **65**, 15621e (1966); 1, 22, 24, 34.
(253) M. C. St. Cyr and W. C. Tanner, *U.S. Dept. Comm., Office Tech. Serv., AD275,432* (1962); *Chem. Abstr.,* **59**, 10301h (1963); 6, 17, 31.
(254) Compagnie de Saint-Gobain, French Pat. 1,473,183 (Mar. 1967); *Chem. Abstr.,* **67**, 76007m (1967); 4, 7, 14, 24, 25.
(255) R. C. Santt, French Pat. 1,468,848 (Feb. 1967); *Chem. Abstr.,* **68**, 70281v (1968); 12, 16, 25.
(256) M. S. Sanzharovskii, Dyl'kov, and P. I. Zubov, *Lakokrasochnye Materialy i ikh Primenenie,* **1964**(6), 17-21; *Chem. Abstr.,* **62**, 8000c (1965); 7, 13, 31, 32.
(257) H. Sato and J. E. Rutzler, Jr., *Am. Chem. Soc., Div. Paint, Plastics, Printing Ink Chem., Preprints,* **20**, No. 1, 124-35 (1960); *Chem. Abstr.,* **57**, 4874f (1962); 1, 5, 32.
(258) K. L. Sayre, U.S. Pat. 2,785,085 (Mar. 1957); *Chem. Abstr.,* **51**, 7059h (1957); 1, 14, 16, 24.
(259) L. C. Scala and D. W. Lewis, German Pat. 1,263,887 (Mar. 1968); *Chem. Abstr.,* **68**, 96538j (1968); 6, 7, 17, 24.
(260) A. M. Schiller, G. A. Tanner, and A. F. Lewis, U.S. Pat. 3,309,261 (Mar. 1967); *Chem. Abstr.,* **66**, 105648h (1967); 2, 12, 13, 24.
(261) H. Schorhorn and L. H. Sharpe, *J. Polymer Sci.,* **A3**, (9) 3087-97 (1965); *Chem. Abstr.,* **63**, 15049d (1965); 1, 12, 16, 32, 33.
(262) H. Schonhorn, *J. Polymer Sci.,* **A1**, 2343-59 (1963); *Chem. Abstr.,* **59**, 6579h (1963); 7, 12, 17, 25, 33, 35.
(263) M. E. Schrader and I. Lerner, U.S. Pat. 3,404,023 (Oct. 1968); *Chem. Abstr.,* **70**, 5027u (1969); 1, 14, 16, 21.
(264) M. Schrader, I. Lerner, and L. Deutsch, *U.S. Dept. Comm., Office Tech. Serv., AD638753* (1966); *Chem. Abstr.,* **67**, 14474c (1967); 14, 21, 32.
(265) M. E. Schrader and I. Lerner, *Nature,* **204**(1960), 759-61 (1964); *Chem. Abstr.,* **62**, 5925b (1965); 14, 21, 32.
(266) M. E. Schrader and I. Lerner, *Proc. Ann. Tech. Conf. SPI Reinforced Plastics Div., 20th, Chicago, 1965,* Sec. 3C; *Chem. Abstr.,* **63**, 11794e (1965); 1, 14, 21, 32.

(267) M. E. Schrader, I. Lerner, and F. J. D'Oria, *Mod. Plastics,* **45**(1) 195-7, 268, 272, 276, 280, 282 (1967); *Chem. Abstr.,* **69**, 11047w (1968); 1, 14, 24, 31, 32.

(268) M. E. Schrader, I. Lerner, F. J. D'Oria, and L. Deutsch, *Proc. Ann. Tech. Conf. SPI Reinforced Plastics Div., 22nd, Washington, D.C.,* **1967**, Sec. 13A; *Chem. Abstr.,* **67**, 74254r (1967); 1, 14, 21, 24, 31, 32.

(269) H. Schultz and C. G. Zike, German Pat. 1,091,748 (Oct. 1960); *Chem. Abstr.,* **55**, 25364g (1961); 1, 24.

(270) K. Schwaner, *Plaste Kautschuk,* **7**, 59 (1960); *Chem. Abstr.,* **54**, 16007f (1960); 2, 12, 24, 31.

(271) T. G. Selin, U.S. Pat. 3,277,047 (Oct. 1966); *Chem. Abstr.,* **66**, 29853d (1967); 6, 13, 14, 15, 16, 17, 24.

(272) J. A. Sergovic, U.S. Pat. 3,328,231 (June 1967); *Chem. Abstr.,* **67**, 74178u (1967); 17, 24, 33.

(273) E. G. Shafrin and W. A. Zisman, *Advan. Chem. Ser.,* **1968**, (87), 20-37; *Chem. Abstr.,* **70**, 39109n (1969); 7, 17, 25.

(274) M. B. Sheratte and J. A. Scott, U.S. Pat. 3,231,425 (Jan. 1966); *Chem. Abstr.,* **64**, 12324h (1966); 1, 11, 12, 13, 17, 25, 33.

(275) G. V. Shiryaeva, G. D. Andreevskaya, and L. V. Belunova, *Vysokomolekul. Soedin., Adgeziya Polimerov, Sb. Statei,* **1963**, 103-6; *Chem. Abstr.,* **60**, 10872h (1964); 4, 14, 22, 24.

(276) V. G. Simpson, French Pat. 1,419,909 (Dec. 1965); *Chem. Abstr.,* **65**, 10761f (1966); 15, 17, 24.

(277) T. R. Smith and A. N. Pines, U.S. Pat. 3,248,358 (April 1966); *Chem. Abstr.,* **65**, 921b (1966); 12, 13, 24.

(278) R. Smith-Johannsen, U.S. Pat. 2,643,964 (June 1953); *Chem. Abstr.,* **48**, 397b (1954); 6, 13, 14, 24.

(279) R. C. Snogren, IAA Accession No. A6638489 (1966); *Chem. Abstr.,* **67**, 5295t (1967); 12, 17, 33.

(280) Societe du Verre Textile, French Pat. 1,519,659 (Apr. 1968); *Chem. Abstr.,* **70**, 97734y (1969); 1, 14, 24, 25.

(281) Societe des Usines Chimiques Rhone-Poulenc, Belgian Pat. 611,687 (June 1962); *Chem. Abstr.,* **57**, 12688b (1962); 2, 24.

(282) V. P. Solomko, V. V. Poletukha, I. A. Uskov, and A. G. Zhigotskii, *Ukr. Khim. Zh.,* **30**(3), 305-8 (1964); *Chem. Abstr.,* **61**, 2005g (1964); 7, 14, 16, 24, 32, 34.

(283) J. L. Speier, Jr., German Pat. 1,090,399 (Oct. 1960); *Chem. Abstr.,* **55**, 27997e (1961); 7, 14, 24.

(284) F. Stastny, R. Breu, and H. H. Koerner, French Pat. 1,507,897 (Dcc. 1967); *Chem. Abstr.*, **70**, 12730w (1969); 1, 14, 17, 24.

(285) Stauffer Chemical Co., Netherlands Pat. Appl. 6,612,420 (Mar. 1967); *Chem. Abstr.*, **67**, 44951y (1967); 14, 17, 24, 31.

(286) W. Stegemann and G. Gunther, German Pat. 952,921 (Nov. 1956); *Chem. Abstr.*, **53**, 14570i (1959); 2, 15, 16, 17, 24.

(287) G. A. Stein and E. R. Pelton, U.S. Pat. 3,246,671 (Apr. 1966); *Chem. Abstr.*, **65**, 9123g (1966); 1, 2, 17, 24, 31.

(288) R. Steinman, U.S. Pat. 2,611,718 (Sept. 1952); *Chem. Abstr.*, **47**, 844d (1953); 7, 14, 22.

(289) R. Steinman, U.S. Pat. 2,688,006 (Aug. 1954); *Chem. Abstr.*, **49**, 4253b (1955); 7, 14, 24, 33.

(290) S. Sterman and J. G. Marsden, *Mod. Plastics*, **40**(11), 125, 127, 129-30, 134, 136, 138, 177 (1963); *Chem. Abstr.*, **59**, 10299d (1963); 7, 14, 24, 34.

(291) S. Sterman and J. G. Marsden, *Mod. Plastics*, **41**(2), 254, 261-2, (1963); *Chem. Abstr.*, **60**, 721h (1964); 14, 17, 24, 32, 34.

(292) S. Sterman and J. G. Marsden, *Plastics Technol.*, **9**(5), 38-41 (1963); *Chem. Abstr.*, **59**, 4113f (1963); 1, 4, 5, 7, 24, 34.

(293) S. Sterman and J. G. Marsden, *Polymer Eng. Sci.*, **6**(2), 97-112 (1966); *Chem. Abstr.*, **64**, 17794d (1966); 7, 24, 34.

(294) S. Sterman and J. G. Marsden, *SPE Tech. Papers*, **1963**, 67-82; *Chem. Abstr.*, **60**, 1883e (1964); 1, 4, 5, 7, 14, 16, 24.

(295) S. Sterman and J. G. Marsden, *Proc. Ann. Tech. Conf. SPI Reinforced Plastics Div., 21st, Chicago, 1966,* Sec. 3A; *Chem. Abstr.*, **66**, 19315r (1967); 7, 14, 24.

(296) S. Sterman and J. G. Marsden, U.S. Pat. 3,259,518 (July 1966); *Chem. Abstr.*, **65**, 15619e (1966); 1, 7, 14, 24.

(297) L. A. Sukhareva, Yu. P. Smirnova, P. I. Zubov, A. V. Otova, and R. Ya. *Khvilivitskii, Plasticheskie Massy,* (10), 31-4 (1965); *Chem. Abstr.*, **64**, 869c (1966); 1, 4, 14, 24.

(298) S. Takada and Y. Muira, Japanese Pat. 5030 (Oct. 1953); *Chem. Abstr.*, **48**, 11113i (1954); 5, 14, 17, 24, 34.

(299) D. Taylor, Jr., and J. E. Rutzler, Jr., *Ind. Eng. Chem.*, **50**, 928-34 (1958); *Chem. Abstr.*, **52**, 16005t (1955); 7, 13, 16, 17, 32.

(300) T. A. Te Grotenhuis, U.S. Pat. 3,249,461 (May 1966); *Chem. Abstr.*, **65**, 2438b (1966); 1, 4, 5, 14, 24.

(301) L. Ter-Minassian-Saraga, *Compt. Rend.*, **260**(22) (Groupe 7), 5770-3 (1965); *Chem. Abstr.*, **63**, 4979h (1965); 32.

(302) E. A. Tertykh, A. A. Chuiko, and I. E. Neimark, *Kolloidn. Zh.*, **28**(2), 278-82 (1966); *Chem. Abstr.*, **65**, 4042a (1966); 24, 32, 34.
(303) K. Thinius and G. Grosse, *Plaste Kautschuk*, **5**, 418-21 (1958); *Chem. Abstr.*, **53**, 9717d (1959); 32.
(304) (a) L. M. Thompson and W. E. Hill, "A Preliminary Evaluation of Silane Coupling Agents as Primers and Additives in Polyurethane Bonding Procedures," NASA TM X-53676; 2, 12, 24, 31. (b) *Ibid.*, unpublished data.
(305) R. L. Tiede, U.S. Pat. 3,253, 948 (May 1966); *Chem. Abstr.*, **65**, 5193d (1966); 5, 14, 24, 31.
(306) S. N. Tolstaya, S. S. Mikhailova, V. M. Luk'yanovich, and E. I. Evko, *Kolloid. Zh.*, **29**(6), 892-3 (1967); *Chem. Abstr.*, **68**, 30558v (1968); 7, 32, 34.
(307) J. W. Trebilcock, U.S. Pat. 3,185,717 (May 1965); *Chem. Abstr.*, **63**, 9515c (1965); 16, 22.
(308) N. M. Trivisonno, L. Lieng-Huang, and S. M. Skinner, *Ind. Eng. Chem.*, **50**, 912-17 (1958); *Chem. Abstr.*, **52**, 15951b (1958); 14, 32, 33.
(309) E. B. Trostyanskava, A. M. Poimanov, and Yu. N. Kazanskii, *Mekhanizm Polimerov, Akad. Nauk Latv. SSR*, **1965**(1), 26-35; *Chem. Abstr.*, **63**, 1939f (1965); 1, 5, 6, 7, 14, 24, 25.
(310) E. B. Trostyanskava, A. M. Poimanov, and Yu. N. Kazanskii, *Plasticheskie Massy*, (8), 20-3 (1964); *Chem. Abstr.*, **61**, 12144g (1964).
(311) E. B. Trostyanskaya and G. M. Gunyaev, *Plasticheskie Massy*, (5) 30-4 (1967); *Chem. Abstr.*, **67**, 54713m (1967); 1, 5, 14, 24.
(312) D. J. Tutas, R. Stromberg, and E. Passaglia, *SPE Trans.*, **4**(4), 256-62 (1964); *Chem. Abstr.*, **62**, 4216a (1965); 14, 24, 32.
(313) L. J. Tyler, U.S. Pat. 3,170,962 (Feb. 1965); *Chem. Abstr.*, **62**, 13388h (1965); 1, 6, 17, 24.
(314) D. A. Uklonskii, E. S. Borodavchenko, V. V. Severnyi, and V. N. Slepich, U.S.S.R. Pat. 193, 066 (19); *Chem. Abstr.*, **68**, 79061f (1968); 6, 24, 31.
(315) N. V. Unilever, French Pat. 1,364,168 (June 1964); *Chem. Abstr.*, **61**, 14901d (1964); 6, 14, 24.
(316) Uniroyal, Inc., Netherlands Appl. 6,610,535 (Feb. 1967); *Chem. Abstr.*, **67**, 100856k (1967); 3, 15, 16, 25.
(317) B. M. Vanderbilt and R. E. Clayton, U.S. Pat. 3,307,967 (Mar. 1967); *Chem. Abstr.*, **66**, 86471b (1967); 3, 14, 15, 24.
(318) B. M. Vanderbilt, U.S. Pat. 3,301,739 (Jan. 1967); *Chem. Abstr.*, **66**, 95896w (1967); 7, 14, 24.

(319) B. M. Vanderbilt and R. E. Clayton, *Ind. Eng. Chem., Prod. Res. Develop.*, **4**(1), 18-22 (1965); *Chem. Abstr.*, **62**, 9330c (1965); 7, 14, 15, 24.
(320) B. M. Vanderbilt and J. J. Jurazelski, *Ind. Eng. Chem., Prod. Res. Develop.*, **1**, 188-94 (1962); *Chem. Abstr.*, **57**, 12711f (1962); 7, 14, 24, 34.
(321) Vereinigung Volkseigener Betriebe Haushalt- und Verpackungs-glas Weisswasser, French Pat. 1,411,565 (Sept. 1965); *Chem. Abstr.*, **65**, 4061a (1966); 1, 14, 17, 24, 31.
(322) Vereinigung Volkseigener Betriebe Haushalt- und Verpackungs-glas Weisswasser, Belgian Pat. 653,776 (Mar. 1965); *Chem. Abstr.*, **65**, 7402g (1966); 1, 14, 17, 24, 31.
(323) R. V. Viventi, H. T. Plant, and R. T. Maker, *Mod. Plastics*, **45**, 289 (1968); 7, 14, 16, 24.
(324) (a) L. M. Vinogradova and A. Ya. Korolev, *Vysokomol. Soedin.*, **B11**(2), 126-9 (1969); *Chem. Abstr.*, **70**, 107562g (1969); 7, 12, 13, 25, 32.
(b) K. N. Vlasova and L. A. Rodivilova, *Plasticheskie Massy*, (2), 19-23 (1960); *Chem. Abstr.*, **54**, 23421i (1960); 7, 12, 13, 15, 17, 31.
(325) G. E. Vogel, O. K. Johannson, F. O. Stark, and R. M. Fleischmann, *Proc. Ann. Tech. Conf. SPI, Reinforced Plastics Div., 22nd, Washington, D. C., 1967,* Sec. 13B; *Chem. Abstr.*, **67**, 33230m (1967); 1, 7, 14, 24, 34.
(326) M. G. Voronkov, E. A. Lasskaya, and A. A. Pashchenko, *Zh. Prikl. Khim.*, **38**(7), 1483-7 (1965); *Chem. Abstr.*, **63**, 11868c (1965); 14, 24, 32.
(327) S. S. Voyutskii, V. G. Raevskii, and S. M. Yagnyatinskaya, *Zh. Vses. Khim. Obshchestva im. D. I. Mendeleeva*, **9**(1), 114-15 (1964); *Chem. Abstr.*, **61**, 807h (1964); 3, 14, 15, 24.
(328) S. S. Voyutskii and Yu. I. Markin., *Vysokomolekul. Soedin.*, **4**, 926-34 (1962). *Chem. Abstr.*, **59**, 1817d (1963); 3, 12, 17, 35.
(329) Wacker-Chemie G.m.b.H., French Pat. 1,361,847 (May 1964); *Chem. Abstr.*, **62**, 5425d (1965); 3, 14, 15, 24.
(330) R. B. Waterhouse and J. H. Schulman, *J. Oil & Colour Chemists' Assoc.*, **38**, 646-51 (1955); *Chem. Abstr.*, **50**, 16267i (1956); 11, 12, 13, 25, 32.
(331) R. F. Wegman, *Adhesives Age*, **10**(1), 20-7 (1967); *Chem. Abstr.*, **66**, 116318b (1967); 1, 12, 22, 32.
(332) H. Weisbart, *Kunststoffe*, **54**(12), 784-90 (1964); *Chem. Abstr.*, **62**, 7474h (1965); 1, 4, 7, 14, 24, 31, 34.

(333) H. Weisbart, *Proc. Ann. Tech. Conf. SPI Reinforced Plastics Div., 20th, Chicago, 1965,* Sec. 3D; *Chem. Abstr.,* **63**, 3989e (1965); 7, 14, 25.
(334) H. Weisbart, E. Behnke, and G. Eckolt, German Pat. 1,022,380 (Jan. 1958); *Chem. Abstr.,* **54**, 11425d (1960); 4, 14, 22.
(335) P. Weiss, *Dechema Monograph.,* **51**(885-894), 99-113 (1964); *Chem. Abstr.,* **62**, 13322f (1965); 1, 16, 32.
(336) T. E. White, *Proc. Ann. Tech. Conf. SPI, Reinforced Plastics Div., 20th, Chicago, 1965,* Sec. 3B; *Chem. Abstr.,* **63**, 7675c (1965); 14, 24, 32.
(337) G. Wiedemann, U. Renner, and B. Ziegenspeck, *Faserforsch. Textiltech.,* **17**, 444-52 (1966); *Chem. Abstr.,* **66**, 38793m (1967); 6, 14, 22, 24, 32.
(338) M. Wismer, V. G. Ammons, and G. W. Miller, French Pat. 1,475,917 (Apr. 1967); *Chem. Abstr.,* **67**, 91439x (1967); 2, 6, 14, 24.
(339) S. E. Wissinger, Jr., U.S. Pat. 3,291,633 (Dec. 1966); *Chem. Abstr.,* **66**, 48931u (1967); 14, 22.
(340) L. L. Yaeger, U.S. Pat. 3,193,429 (July 1965); *Chem. Abstr.,* **63**, 11144c (1965); 7, 14, 24.
(341) L. L. Yaeger, German Pat. 1,034,075 (July 1958); *Chem. Abstr.,* **54**, 26023a (1960); 5, 6, 7, 14, 24, 34.
(342) Yung-Fang Yu Yao, *J. Phys. Chem.,* **68**(1), 101-5 (1964); *Chem. Abstr.,* **60**, 4831h (1964); 13, 23, 32.
(343) P. C. Yates and J. W. Trebilcock, *Reinforced Plastics Div. Soc. Plastics Ind.,* **16**, 8-B (1961); 4, 14, 22, 32.
(344) F. M. Young and L. L. Rouch, *Am. Chem. Soc., Div. Org. Coatings Plast. Chem., Preprints,* **27**(2), 110-18 (1967); *Chem. Abstr.,* **70**, 99910b (1969); 13, 14, 25, 32.
(345) L. N. Zhinkina, L. N. Severnyi, and T. F. Altukhova, *Kremniiorg. Soedin., Tr. Soveshch,* No. 3, 77-81 (1967); *Chem. Abstr.,* **69**, 107312e (1968); 7, 24.
(346) W. A. Zisman, *Ind. Eng. Chem. Prod. Res. Develop.,* **8**(2), 98-110 (1969); 14, 22, 24, 32.

GENERAL BIBLIOGRAPHY OF BOOKS AND REVIEWS

D. J. Alner, *Aspects of Adhesion,* Vol. 1-4, Oxford Univ. Press (Univ. London) London, Annual Review.
H. Alter, *Univ. Microfilms,* Ann Arbor, Mich. Order No. 61-6692, 68 pp. *Dissertation Abstr.,* **22**, 3002 (1962).

C. C. Anderson, *Ind. Eng. Chem.*, **60**(8), 80-7 (1968).
Anon., *Adhesive Age,* 3,(9), 30-4 (1960)
R. E. Baier, E. G. Shafrin, and W. A. Zisman, *Science,* **162**(3860), 1360-8 (1968).
H. Baumann, *Leime und Kontaktkleber; Theoretische Grundlagen, Eigenschaften, Anwendung,* Springer-Verlag, Berlin, 1967.
N. E. Beach, *Plastics Tech. Eval. Center,* Note No. 14, 90 pp., 1966.
J. J. Bikerman, *The Science of Adhesive Joints,* 2nd ed., Academic, New York, 1968.
J. J. Bikerman, *Ind. Eng. Chem.*, **59**(9), 40-4 (1967).
J. J. Bikerman, *Am. Chem. Soc., Div. Polymer Chem. Preprints,* **5**(2), 566-7 (1964).
J. J. Bikerman, *Encycl. Polymer Sci. Technol.*, **1**, 477-82 (1964).
F. Clark, J. E. Rutzler, Jr., and R. L. Savage, *Adhesion and Adhesives,* Society of Chemical Industry, London, 1954.
N. A. De Bruyne, *Plastics (London),* **16**, 308-9 (1951).
R. L. Fiske, *Adhesives Age,* **4**(10), 33-7 (1961).
W. J. Frissel, *Encycl. Polymer Sci. Technol.*, **6**, 740-63 (1967).
M. Fukumura, *Nippon Secchaku Kyokai Shi.*, **3**(2), 90-8 (1967).
E. O. Genzsch, *Gas-Wasserfach.*, **103**, 997-1000 (1962).
Yu. Gorbatkina, *Adhaesion,* **10**(5), 201-6, 208, 210 (1966).
V. R. Gray, *Chem. Ind. (London),* **23**, 969-78 (1965).
T. Hata, *Shikizai Kyokaishi,* **39**(12), 599-609 (1966).
F. B. Hauserman, *Advan. Chem. Ser.*, **23**, 338-56 (1959).
R. Houwink, *Kolloid-Z.*, **151**, 143-7 (1957).
R. Houwink and G. Salomon, "Adhesion and Adhesives," Vol. 1, *Adhesives,* 2nd ed., Elsevier, New York, 1965.
C. Y. Huang, *Kagaku No Ryoiki,* **20**(9), 685-95 (1966).
T. Imoto, *Nippon Gomu Kyokaishi,* **39**(10), 761-99 (1966).
F. Jaeger, *Kautschuk Gummi Kunststoffe,* **18**(3), 155-62 (1965).
M. Joesler, *Materie Plastiche Elastomeri,* **29**(5), 652-60 (1963).
A. Jurecic, *Tappi,* **49**(7), 306-10 (1966).
K. Kanamaru, *Kobunshi,* **17**(197), 740-9 (1968).
D. A. Kardashov, *Tr. Vses. Konf. Kleyam Tekhnol. Skleivaniya, 1st, 1966.* 5-26.
R. E. Keith, M. D. Randall, and D. C. Martin, Spec. Publ. NASA-SP-5085, 1967. Avail. Dep., CFSTI. *Nucl. Sci. Abstr.*, **22**(10), 19517, 1968.
G. M. Kline and F. W. Reinhart, *Mech. Eng.*, **72**, 717-22 (1950).
M. P. Kotov, *Izv. Vysshikh Uchebn. Zavedenii, Tekhnol. Legkoi Prom.*, **6**, 29-36 (1964).

P. S. S. Kumaran, Univ. Microfilms, Ann Arbor, Mich. L. C. Card No. Mic 60-6447, 71 pp. *Dissertation Abstr.*, **21**, 2507 (1961).
H. H. Levine, *Ind. Eng. Chem.*, **54**(3), 22-5 (1962).
J. J. Maheo, *Adhesives Age*, **3**, (4), 35-7 (1960).
H. F. Mark, *Dechema Monograph.*, **51**(885-894), 59-96 (1964).
J. G. Marsden and S. Sterman, *Am. Chem. Soc. Org. Coatings Plastics Chem. Preprints*, **25**(2), 91-100 (1965).
A. Matting and W. Brockmann, *Angew. Chem. Intern. Ed. Engl.*, **7**(8), 598-605 (1968).
Michel, *Adhaesion*, **10**(4), 153-7 (1966).
F. Mittrop, *Tech. Mitt.*, **57**(8), 387-93 (1964).
N. I. Moskvitin, *Physiocochemical Principles of Gluing and Adhesion Processes*, SFCSI-Agr (TT-68-50368), 1969.
R. Mueller, *Tech. Mitt.*, **57**(7), 303-9 (1964).
K. Nakao, *Nippon Secchaku Kyokai Shi.*, **3**(4), 233-44 (1967).
S. Nakazawa, K. Kadowaki, and T. Nakagami, *Senryo To Yakuhin.*, **12**(5), 186-95 (1967).
R. G. Newhall, "Metal-to-Metal Adhesives for the Assembly of Aircraft," Western Business Pubs., San Francisco, Cal., 1955.
M. Okawara, *Nippon Secchaku Kyokai Shi*, **4**(1), 9-19 (1968).
R. S. R. Parker and P. Taylor, *Adhesion and Adhesives*, Macmillan (Pergamon), New York, 1966.
E. P. Plueddemann, *Proc. Ann. Tech. Conf. SPI Reinforced Plastics Div., 20th, Chicago*, 1965. Sec. 19A, 10 pp.
E. P. Plueddemann, *J. Paint Technol.*, **40**(516), 1-9 (1968).
S. Reichel, *Plaste Kautschuk*, **15**(7), 501-4 (1968).
J. E. Rutzler, Jr., *Adhesives Age*, **2**(6), 39-43 (1959); **2**(7), 28-31 (1959).
K. Schnurrbusch, *Plaste Kautschuk*, **4**, 45-7, 53 (1957).
P. O. Seidler, *Adhaesion*, **7**(11), 503-8, 510-12 (1963).
W. Simmler, *Mitt. Ver. Deut. Emailfachleute.*, **13**(11), 77-80 (1965).
L. Stadelmann, *Gummi Asbest*, **13**, 356, 358, 360, 432, 434, 436, 444 (1960).
S. Sterman and J. G. Marsden, *Ind. Eng. Chem.*, **58**(3), 33-7 (1966).
S. Sterman and J. G. Marsden, *Mod. Plastics*, **44-6**, 169-73 (1967).
S. Sterman and J. G. Marsden, *Fundamental Aspects Fiber Reinf. Plast. Compos., Conf., Dayton, Ohio, 1966*, 245-73.
E. Thelen, *J. Appl. Polymer Sci.*, **6**, 150-4 (1962).
R. M. Vasenin, *Adhesives Age*, **8**(5), 18-25; **6**, 30-5 (1965).
S. S. Voyutskii and V. L. Vakula, *J. Appl. Polymer Sci.*, **7**, 475-91 (1963).
S. S. Voyutskii and V. L. Vakula, *Usp. Khim.*, **28**, 701-40 (1959).

W. C. Wake, *Roy. Inst. Chem. (London) Lect. Ser.*,(4)(1966).
A. H. Walter, *Deut. Farben-Z.*, **22**(7), 281-6 (1968).
P. Weiss, *J. Polymer Sci.*, Pt. C, **4**, No. 12, 169-83 (1966).
P. Weiss, Elsevier, Amsterdam, 1962.
W. A. Zisman, *Encycl. Polymer Sci. Technol.*, **1**, 445-77 (1964).

Processing Powdered Polyethylene

A. B. ZIMMERMAN

U.S. Industrial Chemicals Company
Polymer Service Laboratories
Tuscola, Illinois 61953

ABSTRACT

Powdered polyethylenes have been on the plastics processing scene since the early 1960's. In this relatively short period of time, they have captured practically the entire automotive carpet backing market, and are making a significant penetration into the rotational molding market where, potentially, a much larger volume exists. Smaller-volume applications for powdered polyethylenes include the coating of textiles and metals. Very finely ground polyethylene powders are also available for use in specialty applications. Coating and molding techniques and related processing equipment are discussed in this comprehensive report on polyethylene powders.

INTRODUCTION

In the past few years, powdered polyethylenes have gained acceptance in a variety of applications where it is either impossible or inconvenient to utilize the conventional cube or pellet forms. They have achieved recognition as a material offering good design freedom and economy.

Many of the basic problems associated with powdered polyethylenes have been eliminated by the development of more sophisticated processing equipment and improved techniques of handling. Today, these materials are being used in applications where efficiency at high production rates is attained.

Typical application techniques of powdered polyethylenes include molding, coating of fabric, metal and other rigid materials, and specialty applications in which a very finely ground powder is recommended (Table 1).

Table 1. Application Techniques for Polyethylene Powders

Application technique	Processes	Examples
Molding	Predominantly rotational	Large containers, boxes, and other hollow items; drum and other liners; large and small toys, balls, dolls, sporting equipment; hollow display items of all sizes; cases for musical instruments, bowling balls, etc.
Coating fabrics	Carpet backing	Coating and molding automobile carpeting; coating carpets for homes, offices, etc.
	Textile coating	Sealable interlinings, laminated and molded fabrics.
Coating metals and other rigid materials	Dip coating Flood coating Fluidized-bed coating	Coating small parts
	Spray coating Flame-spray coating Electrostatic coating	Coating large metal objects, peelable coatings for corrosion resistance.

Table 1 (continued)

Application technique	Processes	Examples
Specialty applications		Mild abrasive in hand cleaners and scouring ingredient in surgical soaps, lotions, and pastes; mold-release agent and lubricant; chemical carrier and additive for various purposes; filtering medium and pigment-dispersing agent; stiffening agent and binder for fabrics; strip coating and component of strippable latex coatings; bonding agent for other powder materials; various proprietary uses.

The first major breakthrough for powdered polyethylene occurred with the development of a large market in backcoating automotive carpet. The consumption for this application is currently about 30 million lb/yr with substantial increases predicted for the future. Carpet backing represents one of the largest single applications for powdered polyethylene.

A technique offering considerable promise as a large user of powdered polyethylenes is rotational molding. Economically competitive, rotational molding also overcomes some of the limitations of the other molding techniques such as injection or blow and thermoforming. Other powder molding techniques are described in this text, but to date they have had very limited usage.

There are a number of techniques for coating metals and other rigid items using powdered polyethylenes. Small parts can be coated by dipping methods – flooding or fluidized-bed – and large parts usually require a flame-spraying or electrostatic coating technique to provide a protective coating.

The development of very fine particle size polyethylene powders has opened up a new field of specialty applications. Suggested uses for these minutely sized powders include coatings, binding agent, resin additive, lubricant or mold-release agent, paint and lacquer additive, mild scouring agent for cosmetics, and an essential ingredient in many other applications.

RESIN CHOICE

The wide range of applications for powdered polyethylene resins necessitates a careful selection of the correct basic resin properties for each specific use. Normally, the three most important properties to consider are particle size, melt index, and density.

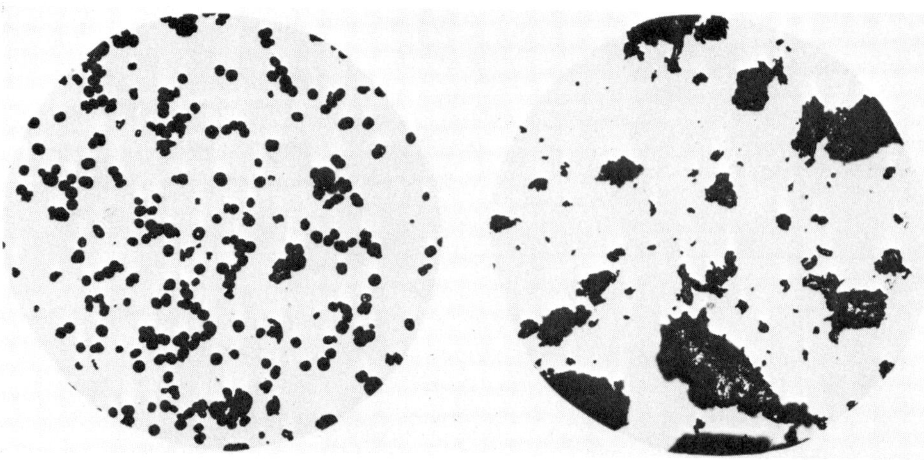

FIG. 1. Photomicrographs showing size and shape of the two basic forms of powdered polyethylene.

Powdered polyethylenes are available in two basic forms (Fig. 1). The mechanically ground powders have a comparatively large average particle size, varying from 300 to 2000 μ (50-10 mesh), and an irregular shape and surface. The very fine powders have an average particle size range from 8 to 30 μ. The particles are sphere-shaped and have smooth surfaces. Details on these very fine powders have been described by Howell (12).

In addition to particle size, melt index and density of powdered polyethylenes are also important properties to consider. The powders are available in melt indexes of 1-70 g/10 min and densities of 0.912-0.960 g/cm^3. Table 2 shows the effects of increases in melt index and density on some significant properties in polyethylene powder applications.

Table 2. How Increases in Melt Index and Density of Polyethylene Powder Affect Processing and End Product Properties

An increase in: Affects following properties	Melt index	Density
Melting point	Decreases	Increases
Flow	Increases	
Impact strength	Decreases	Decreases
Stiffness		Increases
Vicat softening temperature	Decreases	Increases
Resistance to low temperature brittleness	Decreases	Decreases
Barrier properties		Increase

The table indicates that the higher melt index powders are generally more suited for applications where fast melting is desired, such as rotational molding and most coating work. Maximum toughness (impact strength and resistance to low temperature brittleness) of the end product results from the use of lower melt index powders.

Density is an important factor in powder choice primarily because of its effect on stiffness of the end product. Higher density powders have higher stiffness and Vicat softening temperatures. Density also has an important effect on impact strength and resistance to low temperature brittleness. For the best of these two properties, a lower density powder should be chosen. Generally of lesser importance are melting point and barrier properties of the powders, both of which increase with increasing density.

POWDER COATING APPLICATIONS

Polyethylene has a combination of properties which makes it a desirable coating material. It also has certain limitations in coating applications which should be understood to permit proper use of the material. The advantages and limitations are listed in Tables 3 and 4, respectively.

Table 3. Advantages of Polyethylene Coatings

1. Excellent chemical resistance
2. Excellent water resistance
3. Low cost
4. Ease of application
5. Long life (out of direct sunlight)
6. High dielectric strength
7. Good adhesion
8. Excellent mechanical properties
9. FDA approved grades available
10. Heat sealable and heat formable

Table 4. Disadvantages of Polyethylene Coatings

1. Low resistance to permeation for some chemicals
2. Environmental stress cracking
3. Limited color choice for outdoors
4. No air drying formulation
5. Low maximum service temperature

A number of coating techniques using polyethylene powders either as a dry powder or as a dispersion are available. Such factors as convenience, cost, and the nature of the substrate to be coated determine the application technique chosen for any particular substrate.

Table 5. Dry Powder Coating Techniques

Substrates	Coating technique	Examples
Textiles, paper, and other flexible substrates	Feed roll or vibratory feeder	Carpeting, heat sealable fabrics, laminations
Metal and other rigid materials	Spray coating Flame spray Electrostatic spray Floc spray	Tank linings, peelable coatings, protective coatings for large objects
	Dip coating Fluidized bed Electrostatic fluidized bed Nonfluidized bed	Pipe, protective coatings for small parts
	Flood coating Centrifugal and rotational Feed roll or vibratory feeder	Drum linings, pipe

Dry powder coating techniques are usually preferred. The methods are simple and use the lowest cost powders. Table 5 summarizes the various dry powder coating techniques.

Dry powder coatings are generally applied to textiles, paper, and other flexible substrates by a vibratory feeder or a rotating cylindrical feed roll. The coatings are fused with infrared heat and cooled with a chill roll.

Figure 2 is a schematic of one type of equipment used for coating automotive carpeting and some types of textiles. The oven used can be either electric or gas-fired infrared. To coat auto carpet at 60 ft/min an electric

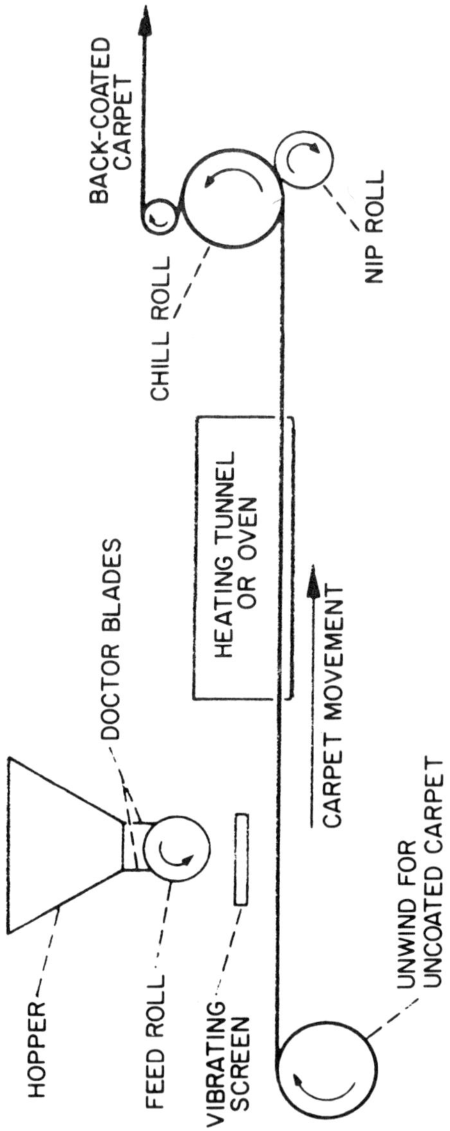

FIG. 2. Schematic of carpet coating line.

oven of about 350 kW will be required based on a coating weight of 10 oz/sq yd. A gas-fired oven of the required size would have an output of approximately 1,210,000 Btu/hr. The chill roll is combined with a nip roll to produce a smooth-surfaced coating. Fabrics can be laminated in-line by applying the second fabric just ahead of the nip and using the nip pressure on the hot coating to produce the bond.

Another type of chill roll setup is shown in Fig. 3. The fabric is not nipped on this type of equipment, producing a rough knobby surface which is desired when lamination is done on a separate piece of equipment.

Following the coating of the carpet, it is cut to the approximate finished size and placed coated side up under infrared heaters. Typical reheat operations place the carpet under the heaters either on a rotating table or a continuous belt. The polyethylene coating is heated to about 400°F for the molding operation. The mold is generally made of fiber glass and may be water cooled or cooled by ambient air temperature (more molds are required when ambient air cooling is used, but the molds are less complicated). The mold is opened and closed pneumatically. The lower portion of the mold is a female section. The jute backing pad is placed in the female section of the mold, and the carpet is inserted with the molten polyethylene side down. As the upper section of the mold closes, the male section produces the draw by forcing the carpet down into the female section where it contacts and adheres to the jute pad. Portions of the female mold not protected by the jute pad are coated with a release agent to prevent sticking of the polyethylene. Following the molding step the carpet is trimmed, "tails" are heat sealed on and the kick pads are applied.

A number of techniques are available for coating metals and rigid substrates. The following steps are generally involved: (a) surface preparation, (b) preheating of the substrate, (c) application of powder, and (d) postheating of the substrate.

Surface preparation ensures adhesion of the coating to the substrate. A mechanical and/or chemical treatment is first used to eliminate corrosion, grease, and oil from the surface. Optimum cleaning conditions must be determined for each type of part. Usually this cleaning treatment alone will produce satisfactory adhesion, but primers can be used if necessary. Surface treatments have been described by Bobee (1), Hmiel (11), and Schonhorn (22).

Primers may serve another purpose, i.e., coverage of difficult areas. For example, USI (unpublished) found that sharp corners on certain steel parts could be covered only when the substrate was immersed in a 2% isopropanol solution of Borden's SO 242F, and then allowed to air dry prior to coating.

Most metal coating techniques require that the part be preheated to a

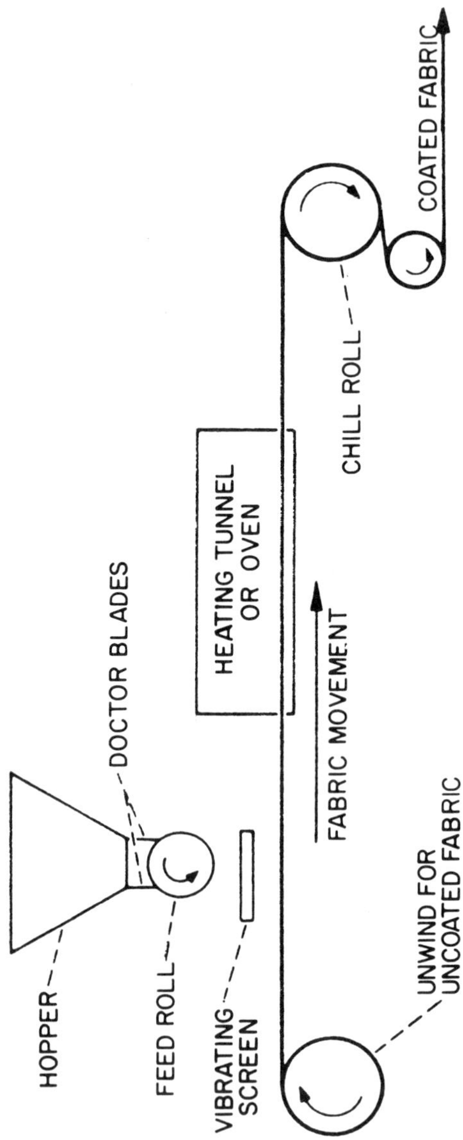

FIG. 3. Schematic of a fabric coating line.

temperature above the softening point of the polyethylene prior to the application of the coating. Electrostatic coating techniques, however, do not require preheat. The optimum preheat temperature must be determined for each part. It is dependent on the mass and material of the part.

Flame spraying, which has been described by Neumann (16), consists of blowing the powder through a flame which partially melts the powder and fuses it as it hits the substrate. The surface is first preheated with the flame to about 400°F. (Only a few square feet are coated at a time so temperature can be controlled.) The flame is then adjusted and the powder passed through the flame and melted on the surface of the substrate. The powder is then shut off and the surface postheated with the flame.

Flame spraying is useful for coating items with a large surface area that could not be heated in an oven. Disadvantages are the problems associated with an open flame and the need for a skilled operator to apply a satisfactory coating.

FIG. 4. Electrostatic spraying apparatus.

Electrostatic spraying offers the advantage that the part need not be preheated since the powder clings to the part by electrostatic attraction. The powder is fed by air pressure through a gun where a high voltage, low amperage charge is applied by an electrostatic generator (Fig. 4). The particles then are propelled to the grounded part where they cling by electrostatic attraction. When the powder builds up to a certain thickness, usually about 8 mils, a dielectric barrier is created and further powder is rejected. The part must then be heated before additional coating can be applied. Davis (6) has discussed electrostatic coaters in more detail.

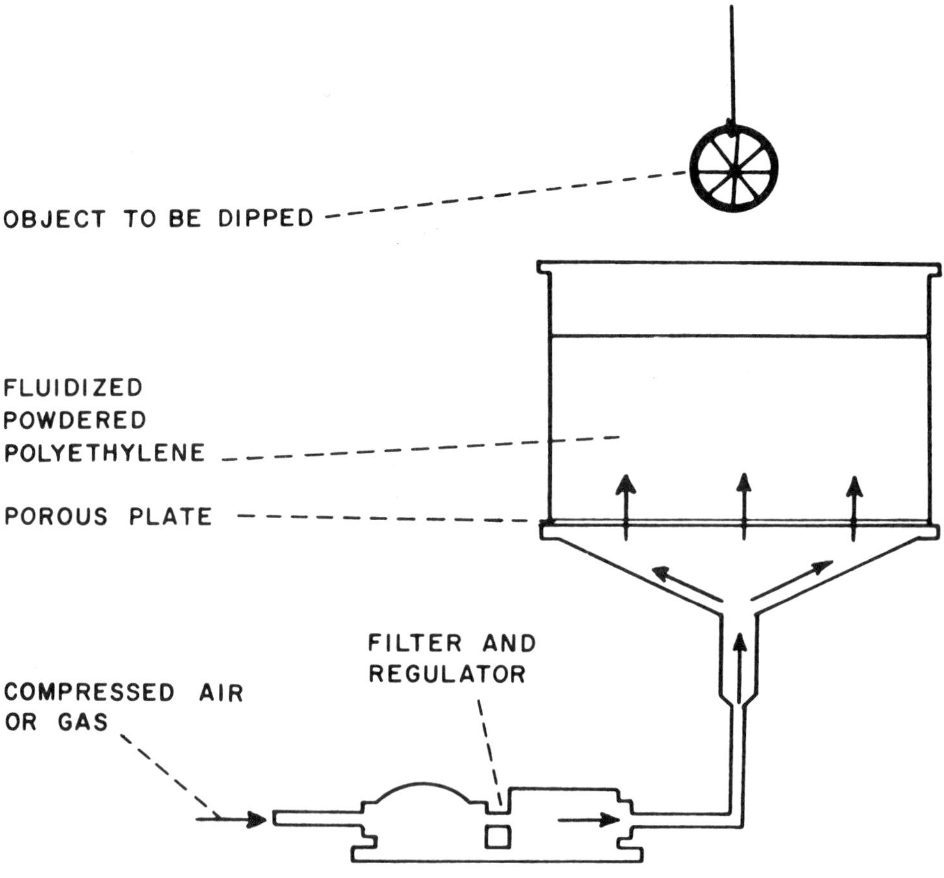

FIG. 5. Schematic of fluidized bed.

Floc spraying is another technique for applying powder to preheated substrates from a specialized gun. Although floc spraying can produce coatings comparable to other methods, a high capacity gun is not available, and considerable operator experience is necessary to obtain satisfactory coatings.

Dip coating involves the dipping of the part into a quantity of powder, usually in the form of a fluidized bed, electrostatic fluidized bed or non-fluidized bed.

The fluidized bed process, U.S. Patent held by the Polymer Corp. (9), involves the agitation of powder in a container by an air stream passing through a porous plate in the bottom of the container (Fig. 5). A cloud of

FIG. 6. Electrostatic fluidized bed.

powder similar to a liquid is formed. The preheated part is dipped into the fluidized powder until the desired coating thickness is obtained. This method is relatively simple, can be used to coat complex shapes, has a high coating rate, and can be adapted to continuous, automated operation. Fluidized-bed coating methods have been reported by Pascoe (18), Nagel and Rickart (15), and Gaynor (8).

The electrostatic coating technique (see Gaynor (8)) has been combined with the fluidized bed to produce an electrostatic fluidized bed (Fig. 6). No preheat is needed with this technique. The powder cloud is charged electrostatically from an electrode in the bottom of the bed which carries a charge from an electrostatic generator.

Parts can also be coated by dipping, rolling, or burying a preheated object in a static bed of powder. This technique is useful where a few parts of small size are to be coated.

Flood coating is the simplest of the dry powder coating techniques. It is best for coating regularly shaped objects where minor variations in coating thickness

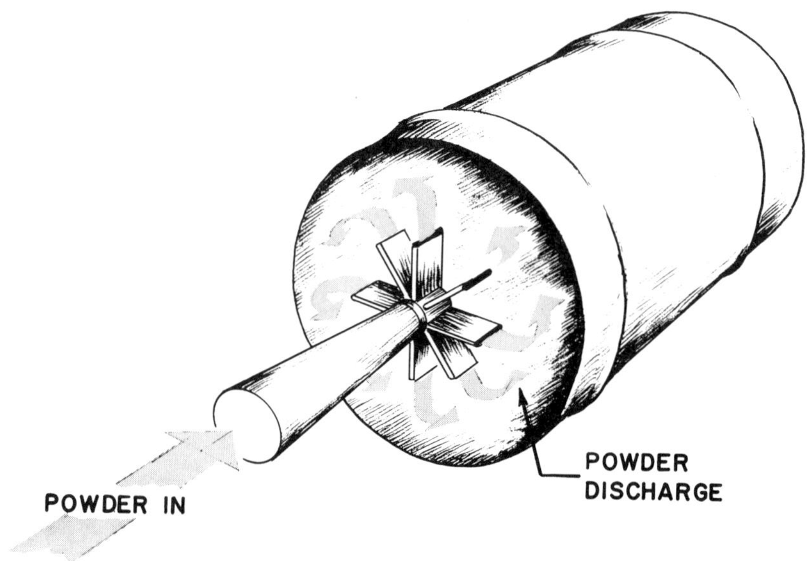

FIG. 7. Schematic of drum and pipe coating equipment.

are not objectionable. The nonfluidized bed is a form of flood coating. Hollow objects can be coated by filling the preheated part with powder and dumping out the excess powder.

In centrifugal or rotational coating methods, powder is placed or metered into the inside of a rotating object and fused (Fig. 7). Linings can be applied to drums or pipes in this manner. This technique has been described by Heisler et al. (10) and Kieffer (14).

The external surface of pipe and similar objects can be coated by feeding powder from a feed roll or vibratory-type feeder (Fig. 8) while the preheated part is being rotated. Flat sheets or plates could also be coated on a continuous conveyor with this type of feeding equipment, possibly using a chill roll-nip roll to smooth the coating similar to the technique previously described for carpet and fabrics.

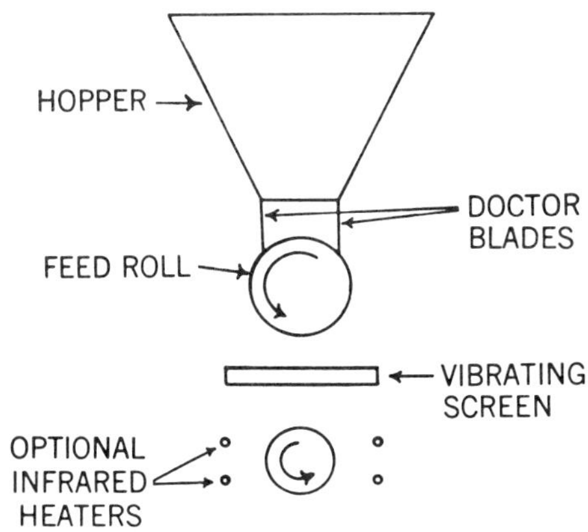

FIG. 8. Schematic of pipe coating equipment.

Following the application of the coating it is usually necessary to postheat to fuse the coating unless the part is massive and retains heat. Postheating is generally carried out at temperatures of 350°F or higher. Many types of heating equipment can be used. Infrared is excellent where shapes are not complex. Hot air heating can also be used provided the air velocity is not high enough to disturb the powder. The proper postheat may oxidize the surface just enough to improve scratch and abrasion resistance.

DISPERSION COATING TECHNIQUES

Dispersions are prepared by mixing the powder with a suitable liquid. An organic dispersion can be prepared by mixing the powder into a suitable organic liquid or combination of liquids. To obtain a stable dispersion, the density of the organic liquid should be close to that of the powder. A suitable solvent for dispersions is a mixture of 31% trichloroethylene by weight and 69% methanol or ethanol. Viscosity of the dispersion can be varied by varying the solids content (maximum solids content is about 45% for organic solvents).

Most of the information on dispersions applies specifically to low and medium density polyethylenes. Hmiel (11) describes stable high density polyethylene dispersions and a method of coating.

Aqueous dispersions require a wetting agent, preferably a nonionic surfactant unless ionic surfactants are required for some specific reason. A minimum surfactant level of 0.2% based on the total weight of the system is recommended. Aqueous dispersions are often prepared by making a paste with the surfactant and resin in about half of the water, and then diluting with the balance of the water and mixing in a high shear mixer. Up to 70% solids contents have been prepared. Water dispersions offer the advantage that flammable and expensive solvents are not used. Water dispersions are best for coating nonferrous items, however.

If air bubbles are present in the dispersion, pinholes will be formed in the coating. Care must be taken not to mix air into the dispersion, and a low foaming surfactant may be used for water dispersions. Such bubbles can be removed by deaerating the dispersion with ultrasonic or vacuum deaerating equipment. Dispersion coating techniques are shown in Table 6.

Several techniques for coating fabrics and other flexible substrates with dispersions are available. Following the coating application the coating is fused, cooled, and nipped if a smooth surface is desired.

Roller coating is a convenient technique for applying dispersions to flat surfaces. A wide variety of equipment can be used. The optimum dispersion viscosity should be determined for each application, as it depends on the speed and coating thickness desired.

Coatings can also be applied on doctor knife equipment. Again, viscosity is a critical factor.

Polyethylene powder dispersions can also be applied with conventional paint spray equipment. Choice of spray equipment depends on the output and width of the spray pattern desired. Atomization pressure in the gun should be held to a minimum to prevent blowing the powder from the surface.

Table 6. Wet Coating Techniques with Powder Dispersions and Pastes

Substrates	Coating techniques	Examples
Textiles, paper, and other flexible substrates	Roller coating Doctor knife coating Spray coating	Specialty paper coating, nonwoven fabric
Metals and other rigid materials	Dip coating Spray coating Slush coating	Protective coatings Drum linings

Polyethylene powder dispersions can also be applied with "airless" spraying units. In this case the gun screen must be left out because it will clog and stop the spray action.

Automatic spraying equipment suitable for use in continuous production lines is available.

Several techniques are also available for coating metal and rigid substrates. After surface preparation like that preceding dry powder coating, the coating is applied. On vertical surfaces the wet coating may run down and produce a thicker coating at the bottom than at the top. Inverting the object after application of the coating may help with this problem. Following the application step, the part is fused to drive off the unevaporated liquid and form the coating.

Dispersions can be used to dip coat objects which have an irregular shape making them difficult to coat by other means. The object is dipped into the dispersion to apply the coating (Fig. 9). Dip time can be as little as 1 sec for thicknesses of up to 7 mils.

Slush coating techniques can be used to coat the inside of hollow objects such as drums. The dispersion is poured into the container and the container rotated in order to wet all surfaces. The excess dispersion is poured out and the coating is fused. More than one coating cycle may be necessary to get a pinhole-free coating on the inside of a container.

ROTATIONAL MOLDING

In 1961, the first polyolefin powder, polyethylene, was publicly demonstrated to the rotational molding industry. The industry quickly realized that

FIG. 9. Dip coating using a solvent dispersion.

polyolefin powders offered their processing technique new applications in the plastics field, many of which would be virtually impossible to fabricate by any other method. Rotational molding of polyolefin powders is distinctively different from other plastics molding processes such as blow and injection molding in that it (a) uses resin powder instead of resin pellets; (b) melts the powder inside the mold instead of forcing molten resin, under pressure, into the mold; (c) uses biaxial rotation of the mold; and (d) utilizes molds that are comparatively inexpensive because of their simplicity owing to the lack of pressure in processing and lack of coring for water cooling. Rotational molding techniques involving polyolefin powders have been developed so that in the production of many items they can compete with blow and injection molding and thermoforming.

The advantages of rotational molding when compared with the other molding techniques are summarized in Table 7.

The selection of the proper polyethylene powder is essential in rotational molding to obtain good end products. Specifically, the high temperatures used in this processing technique bring about a risk in chemical degradation. Therefore, the effect of melt index and density of powdered polyethylene on the end product properties and processability are quite critical for rotational molding, while those of particle size are considerably less.

Table 7. Advantages of Rotational Molding

1.	Costs for molds and tooling are extremely low, especially for short runs.
2.	Prototype molds can be manufactured for experimental purposes without great expense.
3.	In most cases, secondary tooling is eliminated.
4.	Little or no waste scrap – the desired weight of the part is placed into the mold.
5.	Dimensional tolerances of the side walls and bottom of the moldings are easily controllable.
6.	Pieces with undercuts and intricate contours can be molded easily.
7.	Minimum of cross-sectional deformation and warpage.
8.	Relatively stress-free items are produced because of a minimum of locked-in stresses.
9.	Identical or similar items or different sections of one piece can be molded at the same time in different colors on a single spindle.
10.	Plastic or metal inserts can often be molded as integral parts of the item.
11.	Double-wall constructions are feasible.

Higher melt index polyethylenes are generally more suited for applications where high production rates are a primary requisite such as rotational molding of low and medium density powders. However, maximum toughness (impact strength as well as resistance to low temperature brittleness) of the end product results from the use of lower melt index powders. Thus, lower melt index powders are more suited for rotational molding of linear polyethylene powders; a lower melt index increases toughness of higher density resins.

Generally speaking, low density resins are preferable whenever stiffness is not essential or is undesirable, as for many toys, and only light loads are to be expected. Medium density resins are useful for self-supporting items which require higher heat distortion resistance or stiffness that low density resins will

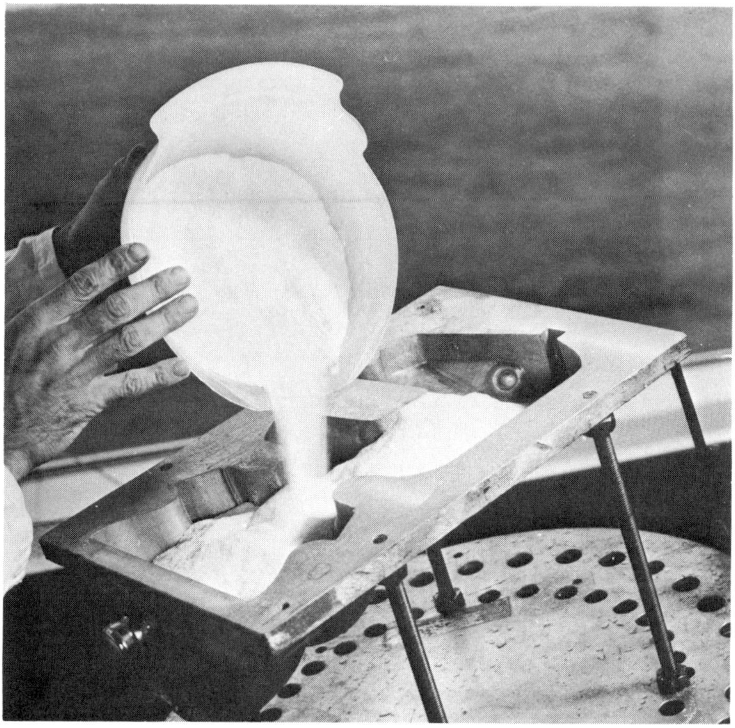

FIG. 10. Filling a mold with a predetermined weight of powder.

not provide. High density resins impart highest rigidity to the end product, which frequently permits reduction of the wall thickness and therewith of the cost per piece.

There are six basic steps associated with the powder molding of a finished product. They are:

(1) The cavity, or multiple cavities, in the bottom parts of the unheated molds are charged with a predetermined weight of powder desired in the end product. Figure 10 shows this step.

(2) The mold parts are clamped in place. For totally enclosed pieces, the entire mold is made of heat conductive material. When one or both ends of the pieces are open, heat insulating covers are used to close the mold.

FIG. 11. Heating the charged mold in an oven while rotating the mold around two axes.

(3) The charged molds are placed in an oven where they are heated (Fig. 11) while simultaneously rotated around two axes in planes at right angles to each other. Figure 12 shows the principle of such biaxial rotation schematically.

(4) The double revolving motion results in forming hollow objects in every mold cavity, the powder being evenly distributed to form walls of uniform

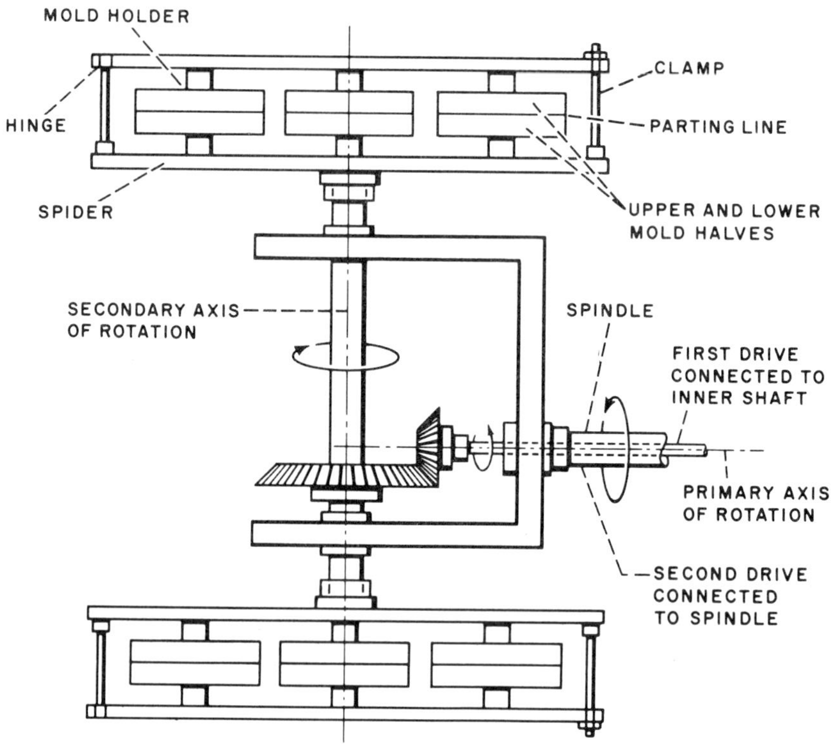

FIG. 12. Schematic presentation of biaxial rotation. The spindle is turned on the primary axis while the molds are rotated on the secondary axis. Usually the ratio of the two simultaneous rotations is determined by the gear, which may be exchangeable, or by two motor drives.

thickness when the resin fuses (except where heat insulating covers are used). Weight and wall thickness of the molded items can be modified by increasing the amount of powder initially put into each lower mold cavity.

(5) When all the powder has fused into a homogeneous layer on the walls

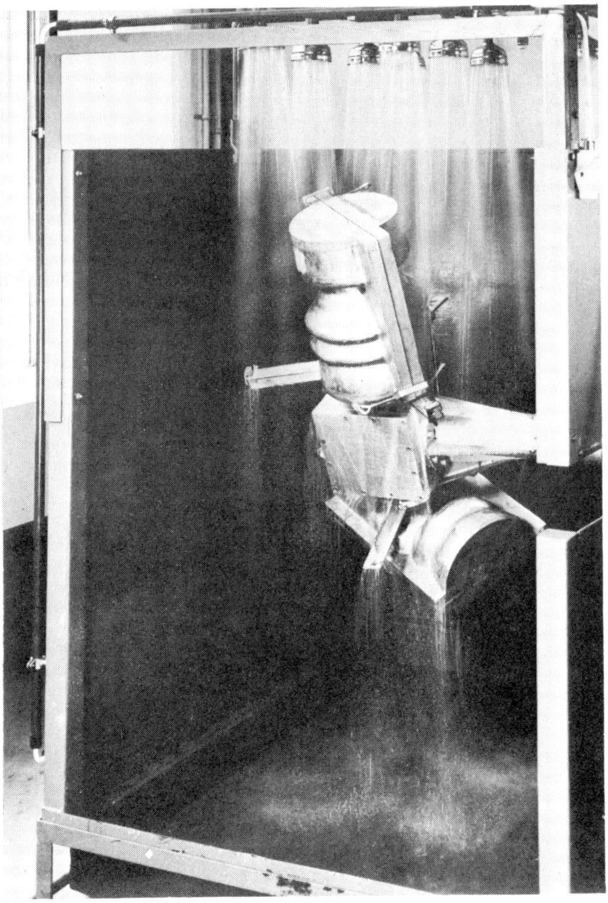

FIG. 13. Cooling the mold while it is still being rotated.

of the cavities, the mold is cooled while still being rotated (Fig. 13).

(6) The mold is opened (Fig. 14) and the molding is removed (Fig. 15). Then the mold is readied for the next cycle.

A mold required for rotational molding can be simple because very little pressure is exerted in the process and no coring for water cooling is necessary. Therefore, such a mold costs only a fraction of what a comparable injection or

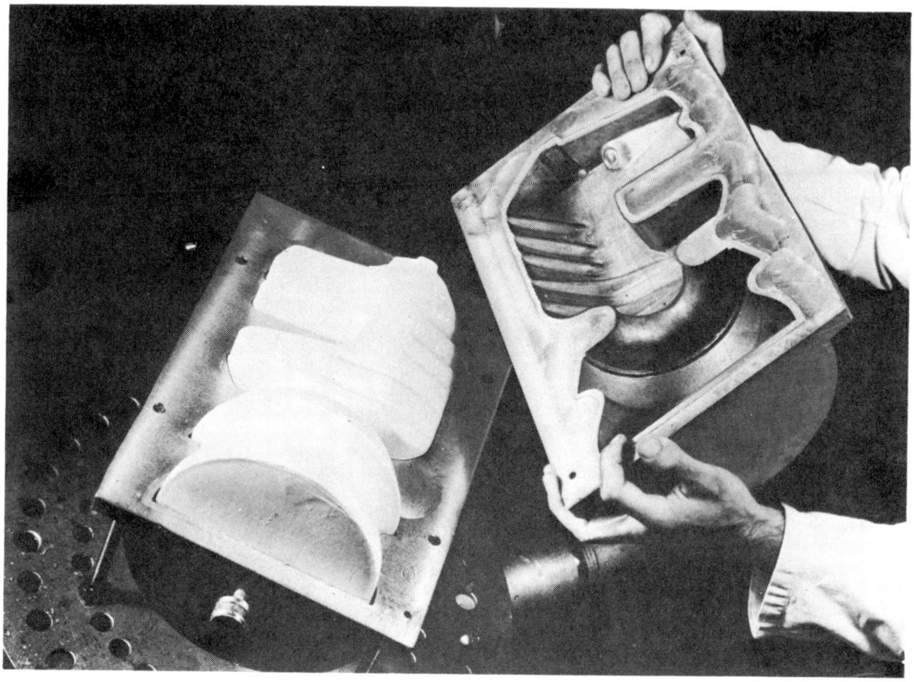

FIG. 14. Opening the mold.

blow mold would cost. Split molds are required for piece removal when rotationally molding powders. The selection of rotational molds depends mainly on the size, shape, and surface finish of the piece to be molded and also on whether one or several molds are to be made for a particular piece. Molds should be as thin walled and light weight as practical.

Possible materials of construction are sheet steel, cast aluminum, or electroformed copper-nickel. Aluminum or electroformed copper-nickel are recommended for small- or intermediate-sized molds, complex shapes, and when several molds of one design are required. Steel should be used for most large molds and for a prototype mold of a single design, because it is the most economical material.

The type of equipment used for rotational molding determines the wall thickness of the mold. Molds used in a hot liquid machine should have a

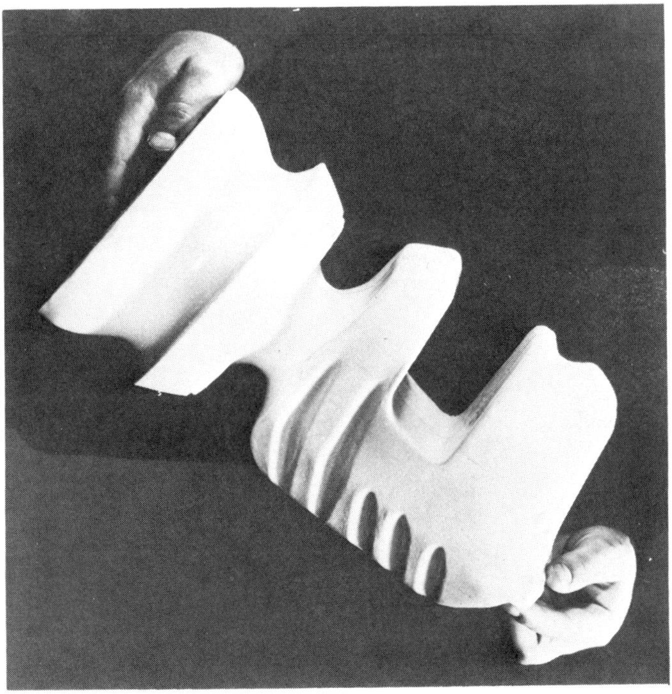

FIG. 15. Removing the molded part.

greater wall thickness (about 1/2 in.) than molds to be used in a hot air convection oven, which should be about 1/4 in. thick. Thicker mold walls permit the high temperatures caused by the fast heat transfer of the liquid to heat the powder uniformly without causing hot spots and with less risk of mold warpage. Thinner mold walls will result in a cycle reduction in a convection oven. Steel molds should be constructed of 10-18-gauge (3.3-1.2 mm) steel, whenever possible. They should have flat steel flanges (see Fig. 16) which should be stress-relieved and machine-matched after they are welded to the mold.

Any of the three types of flange-mating surfaces shown in Fig. 16 are suitable. The flat type can be used on either convection oven or hot liquid machines. It simplifies mold maintenance but has a higher initial cost than either the offset or tongue-and-groove surface. The mating surfaces should be

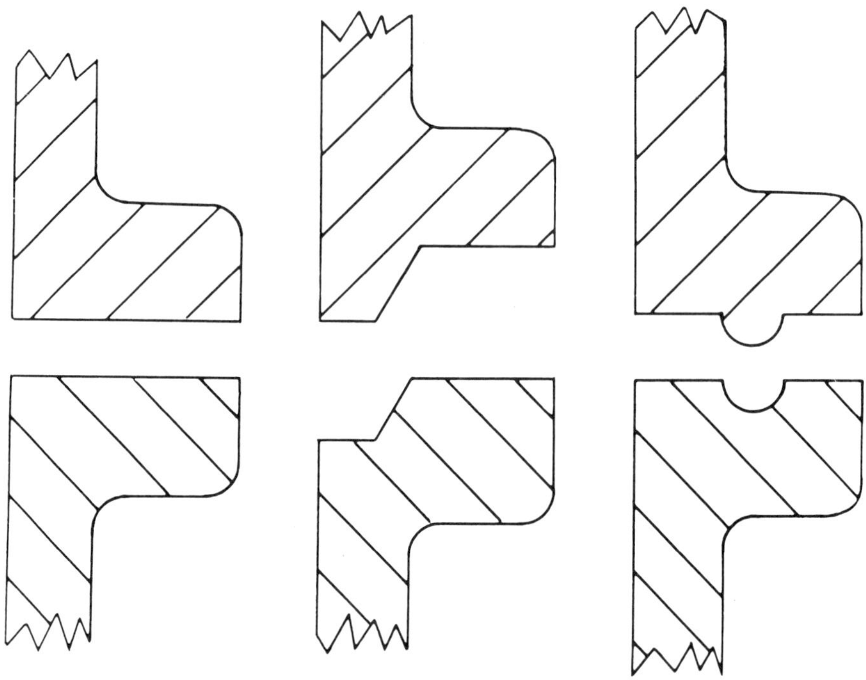

FIG. 16. Three rotational mold flange-mating surface types (left to right): flat, offset, tongue and groove.

machined smooth for a good fit, and the molds should be stress-relieved before the parting lines are matched. Hinges on the parting line of an aluminum mold should be avoided.

When openings are desired in rotationally molded pieces, insulating lids or inserts can be used. Any type of insulating material can be used which keeps the powder from fusing at the area over which it is applied. Teflon, asbestos board, and silicone foams are commonly used. If thin-walled sections are desired in a molded piece, they may be obtained by covering a section of the

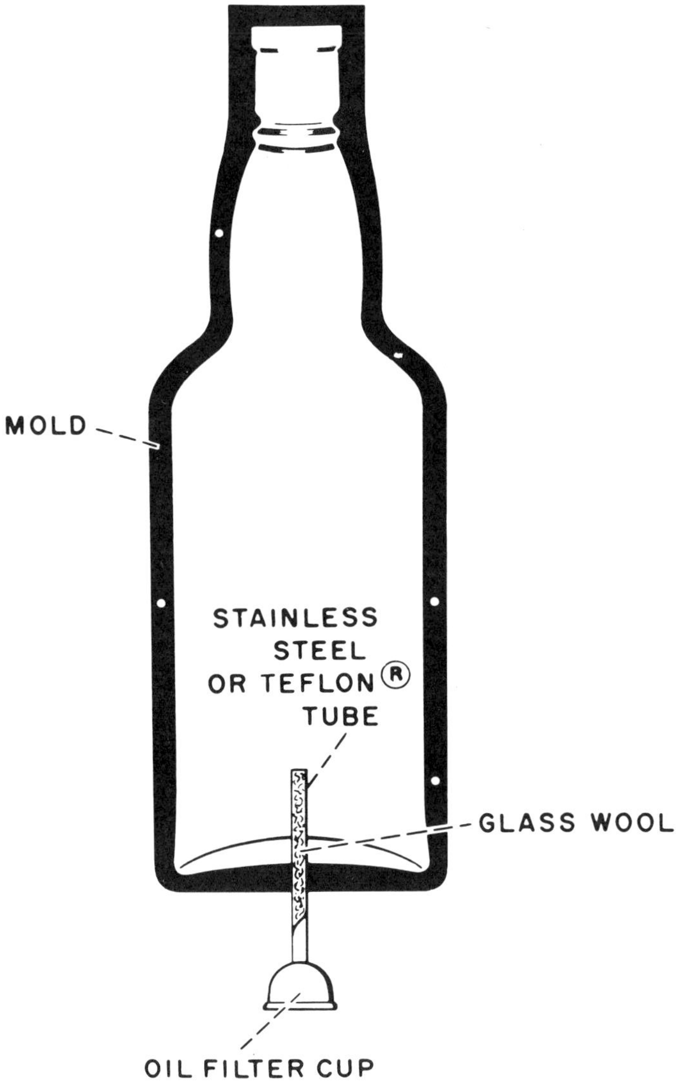

FIG. 17. Schematic of a vented rotational mold for a display bottle.

mold with an insulating material which will result in a smaller amount of material sticking to it at that point. Another method of influencing wall thickness at selected areas of the molded item (when infrared mold heating is used) is by painting. Dark paints on the outer mold surface will result in higher heat absorption and, consequently, thicker walls in the molded part. Light-colored paints such as aluminum will reflect the heat and result in thinner walls. Paints used for this purpose must be thermally stable at the high oven temperatures used. Paints have a relatively small effect on wall thickness compared with insulating covers made of materials such as asbestos.

Venting of rotational molds is often recommended to maintain atmospheric pressure inside the closed mold during the entire molding cycle. A vent will reduce flash and piece or mold distortion and lower the pressure needed to keep the mold closed. It will also prevent blowouts caused by pressure. Finally, it will permit the use of thinner walled molds. Figure 17 is a schematic presentation of a vented mold. The vent is a thin-walled metal or Teflon tube which extends to near the center of the mold. It must enter the mold at a point where the opening which it will leave will not harm the appearance or utility of the molded item. The vent is filled with a material such as glass wool to keep the powder charge from entering the vent during rotation. The end of the vent outside the mold should be protected so that no water will enter during cooling or so that no hot liquid will enter during heating if that type of heating system is being used.

To obtain satisfactory mold release of powdered polyethylene parts it is recommended that the mold cavities be kept clean and a proper mold release agent be used. Most foreign material found in mold cavities can be removed with an abrasive cloth such as emery paper. Molds which have previously been used with vinyl or other plastic material may require special cleaning. The most effective way to remove such material from the mold surface is sand blasting, using aluminum oxide rather than sand. This procedure will result in a matte-finished surface. Once the cavities are clean, one of the semi-permanent fluid mold release agents can be applied to aid in part removal.

The mold handling system includes all the equipment required to transport the mold (or molds) through the loading, heating, cooling, and unloading steps and to bring about the biaxial rotation typical for rotational molding. The system's main parts are the spindle, the mold mount (spider), and the drive (see Fig. 18). A machine can have one or several spindles. With the single-spindle machine such as the one shown in Fig. 18 the production cycle time is the sum of the intervals required to complete all the processing steps – mold charging, heating, cooling, and part removal. The single-spindle

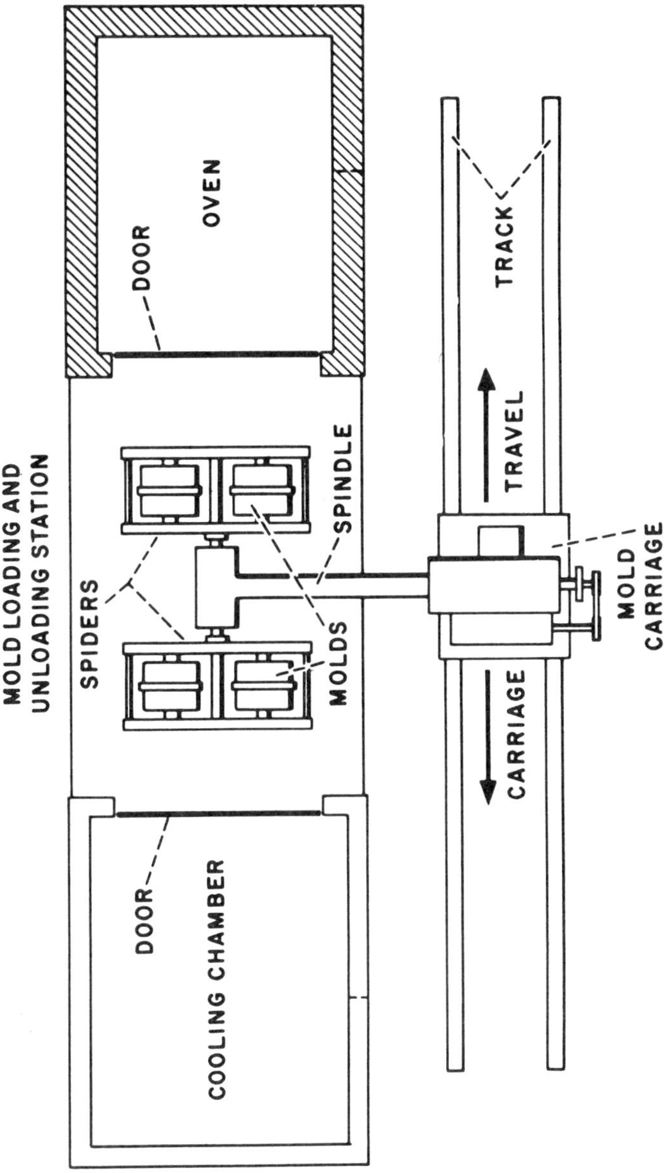

FIG. 18. Schematic of a single-spindle rotational molding machine.

is suitable for molding prototype pieces and for small production runs.

By using a multiple-spindle or carousel-type machine (Fig. 19) production rates can be increased. Multiple-spindle machines are usually wheel-shaped. The spindles, each carrying a group of molds on a single large mold, are mounted on a common hub. The wheel is indexed horizontally from station to station.

Figure 20 shows a comparable three-spindle rotational molding machine. The spindle with the multicavity molds at left is about to enter the oven. The spindle with the large single-cavity mold is just emerging from the cooling station at the right. The spiders with the molds are moved clockwise, first through the gas-heated oven and then through the water-spray cooling unit in a circular path. The three-station unit is a much used multiple-spindle machine. It is available with either hot air or liquid spray heating. While one spindle is

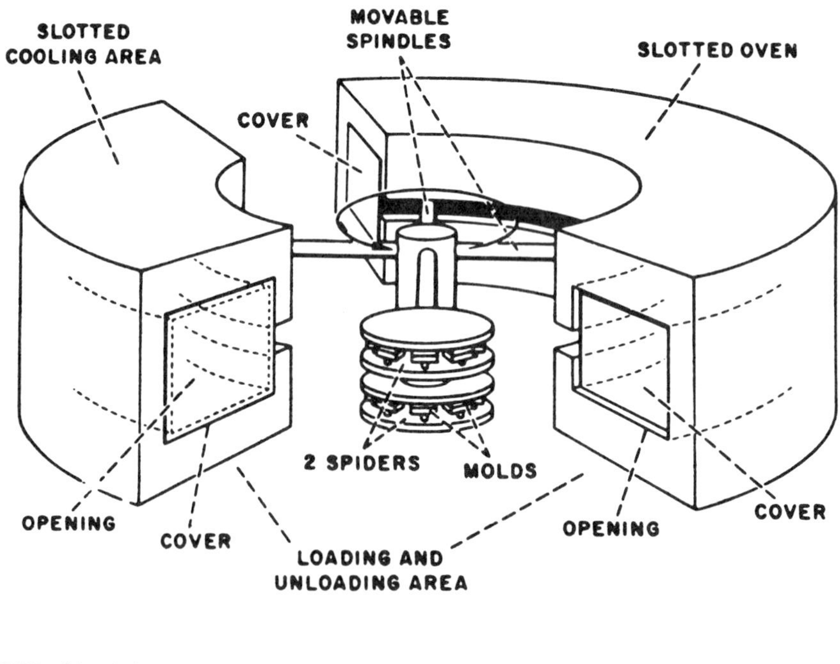

FIG. 19. Schematic of a typical rotary, multiple-spindle (in this case a four-spindle) mold handling system.

in the oven, a second spindle is being cooled. In the third station, both part removal and resin loading take place.

If one spindle and one station of a rotary unit are used for each operation, production time will be roughly that needed to complete the longest processing step. Frequently, two operations (for instance, piece removal and mold loading) can be combined into one station, thus reducing the number of spindles required. However, the addition of more spindles permits more time for processing steps which take longer. This increases production rates. For instance, the molding unit can be designed so that at least two spindles are in the heating unit at all times. This permits more time for heating if this is the longest part of the cycle (which is usually the case).

FIG. 20. A three-spindle rotational molding machine. Photo courtesy of McNeil Corp., Akron, Ohio.

Mold handling systems vary greatly in the way they are used. In the design shown schematically in Fig. 18, the spindle and drive are mounted on a carriage, which moves on tracks between three positions. Following the heating step, the carriage and molds are moved to the cooling chamber and then to the loading and unloading station.

In the pivoting-arm design, the spindle is attached to a pivot instead of being mounted on a carriage. The arm swings in a 90° arc when moving from the oven to the cooling chamber (Fig. 21). As indicated in the figure, a second

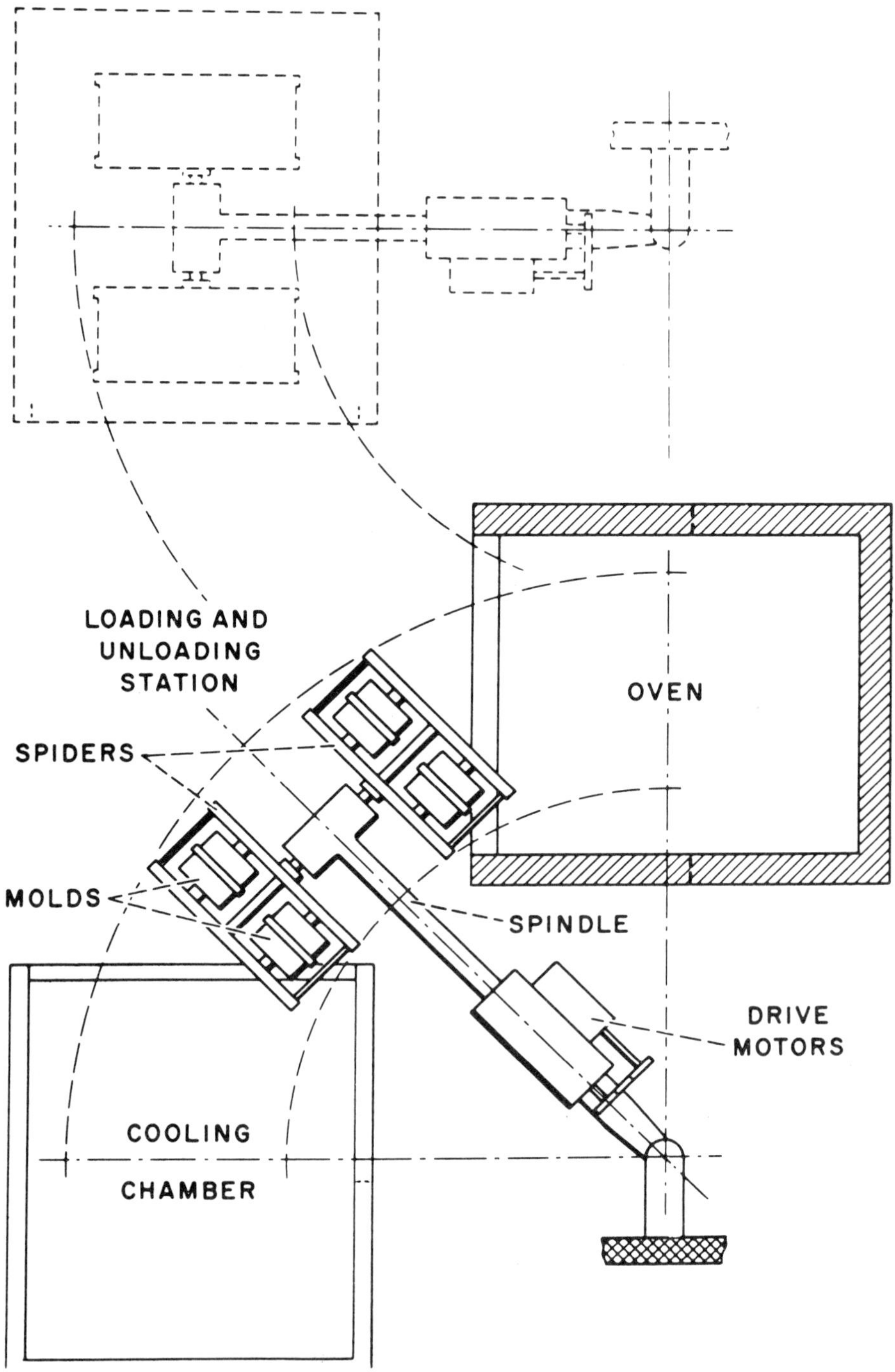

FIG. 21. Schematic of a pivoting-arm mold-moving system.

pivot spindle and cooling chamber can be added. This doubles the production rate without requiring another oven.

To reduce tooling costs for low volume production, different parts may be run on each spindle. If each part requires similarly long heating or cooling periods, only individual speed controls for each spindle are required. If heating or cooling periods differ markedly between parts, individual cycle controls for each spindle are also necessary. This reduces efficiency.

In the fixed-spindle rotational molding machine (Fig. 22), which has been in use for a number of years in vinyl plastisol molding, a roller-top table is used to transfer the mold assembly between the three processing stations. Two drive systems, one in the heating, another in the cooling chamber, are required to rotate the molds.

Figure 23 illustrates two ways of mounting molds on the spindle. The most common method is to attach the mold (or several molds) to platforms (spiders) lying directly above and below the spindle (double-centerline mounting). This mounting is preferred for the production of small items because it permits easier mold loading and unloading. For molding large items, the so-called offset arm is preferable. This way of mold mounting makes possible the rotation of much larger parts. Moreover, it minimizes variations in wall thickness by bringing the centroid (center of volume) of the mold close to the intersection of the primary and secondary axes of rotation. A wide difference in peripheral velocity from top to bottom, caused by a greatly offset mold, may be one reason for poor material distribution and nonuniform wall thickness of the moldings.

A variety of drive systems is available to provide biaxial rotation of the molds. They include either a variable speed drive or constant speed drive with interchangeable gears or chain sprockets.

Most rotational molding is carried out with a 10-12-rpm rotation of the primary axis and a 2-3-rpm rotation of the secondary axis, that is, an average ratio of 4 or 5 to 1. For many applications, this ratio has proved satisfactory. Provisions should, however, be made for changing the ratio to correct nonuniform powder distribution and wall thickness which can occur at the average rotational ratio. If many different parts are expected to be produced on one machine, variable speed drives are recommended to eliminate the need to change gears or sprockets when changing speed or ratio of rotation. If the drive allows for variable speed rotation of each axis independently over a range of 1-25 rpm, any practical rotational ratio can be obtained.

The three most commonly used methods by which molds are heated are hot air convection, hot liquid conduction, and infrared radiation (either gas-fired or electrical heaters). Choice of the most suitable type of heat source

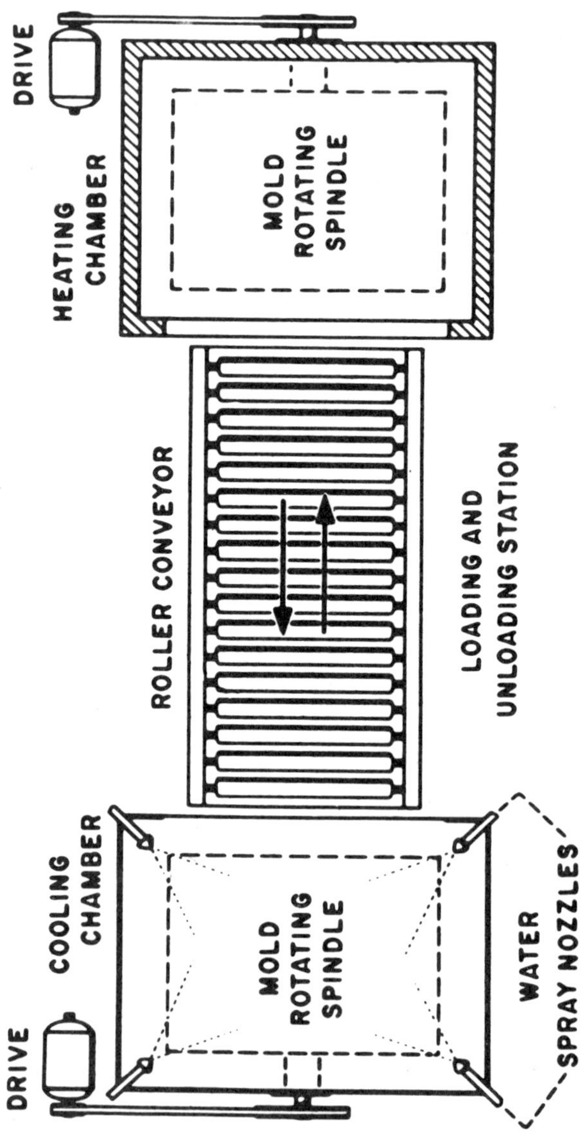

FIG. 22. Schematic of a fixed-spindle mold-moving unit.

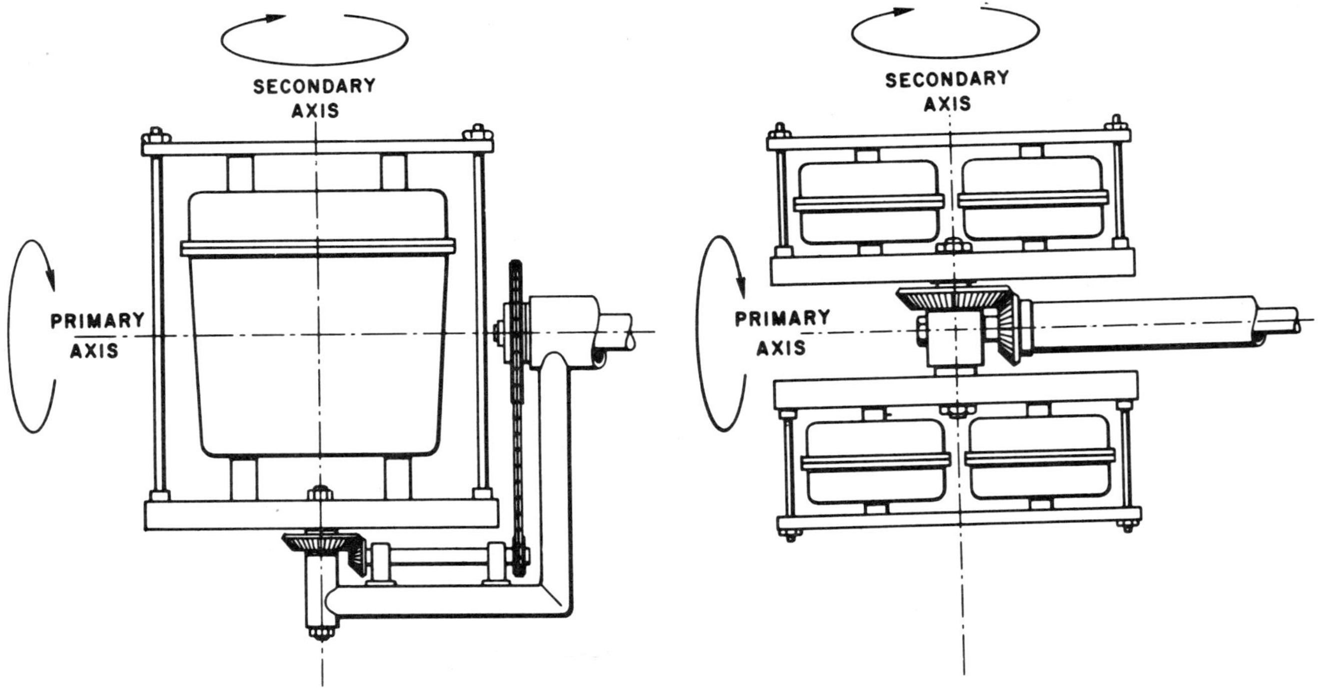

FIG. 23. Schematic of mold mounting systems. Left: offset arm mounting; right: double center-line mounting.

depends mainly on the piece geometry, number of molds on each spider, and required production rate.

Normal oven temperatures are 400°-700°F (205°-370°C) and, frequently, as high as 900°F (480°C). This high temperature will usually not damage thin-walled moldings, but it may damage some molds. Ovens must be well insulated to minimize heat losses.

Hot air is the most commonly used heat source for rotational molding (see Fig. 20). This heat source is particularly suitable for molding thin-walled pieces [less than 60 mils (1 1/2 mm)]. Oven temperatures may vary from 400°-900°F (205°-480°C). The higher the oven temperature, in this range, the shorter the heating cycle.

Accurately directed air flow is important. This can be obtained by baffling the air blower. Equally important are the velocity of the air and uniformity of the temperature throughout the oven. A high velocity scrubbing action is needed for maximum heat transfer. Hot spots in the oven must be avoided when items of uniform wall thickness are to be molded. Excess heat capacity of the oven is desirable so that the desired operating temperature can be rapidly attained after insertion of the cool, charged mold.

Gas-fired, oil-fired, or electrically heated hot air ovens are available. If temperatures are frequently changed, a recording temperature controller is recommended.

Heating by means of hot fluid sprays is also in use. In such ovens, a non-corrosive, inorganic salt is the heat transfer medium. Molten wax and high temperature heat transfer oils have been used, though less successfully.

Operating temperatures in the liquid spray system are normally in the range of 450°-550°F (230°-290°C). The heat transfer liquid is brought up to operating temperature in a melting tank, or reservoir, containing gas fire tubes or electrical immersion heaters. A heating period of up to 2 1/2 hr, depending upon the volume of heating medium used, is required to reach operating temperature. The fluid is pumped through a network of spray nozzles onto the rotating mold assembly. It is important that the melting tanks have enough capacity to maintain the operating temperature throughout the heating cycle. The temperature drop in the melting tank should not exceed 10°-15°F (5 1/2°-8 1/2°C) during the mold heating cycle.

In this heating system, heat is transferred by conduction rather than convection. Hot liquid spraying, therefore, normally heats the molds faster than hot air. It is generally better suited for molding pieces with heavy walls and for the production of more complex shapes.

The molds are usually rinsed with water, to remove the heating fluid, before leaving the oven. Since unrecoverable heating fluid can mean the difference between profit and loss in hot liquid spray heating, it is essential that such equipment have a good system to recover and reuse the heating medium.

Infrared heating with either gas-fired or electrical radiation heaters is mainly used on special-purpose and prototype rotational molding equipment.

Heating with infrared is very fast and efficient, but this heat source is limited to molding simple shapes and single molds. Multiple- or complex-shaped molds cannot be infrared-heated because the radiant heat cannot strike all parts of the mold uniformly. Shielded areas of the mold will not attain as high a temperature as exposed parts of the mold surface. Consequently, thin spots will occur in the wall of the molded item.

Spraying with water is the most common method of cooling the mold. During the cooling process the mold should be rotated, for failure to do so will result in the soft molten polymer sagging toward the bottom of the mold. A highly atomized water spray results in uniform and rapid cooling. Blowers or compressed-air nozzles in the cooling chamber increase flexibility of the cooling system. For critical cooling applications, air and water can be used intermittently.

Since quick cooling makes the over-all molding cycle shorter, it is desirable for economic reasons. However, if cooling is too rapid, warpage of the molded piece may occur. If this happens, the cooling rate should be slowed by raising the temperature of the cooling bath or by using forced air in combination with the water spray.

Molds may be filled manually or automatically. For manual filling, the required quantities of powder are premeasured, by weight or volume, in a container and then dumped into the lower halves of the molds as shown in Fig. 10. Manual filling is advantageous when filling differently colored items simultaneously in a multiple-cavity mold.

For long runs, automatic mold-filling equipment is useful. The bulk density – measured in lb/cu ft (1 lb/cu ft = 16 g/liter) – of powdered polyolefins may vary, and to avoid differences in wall thickness between pieces, the powder charge should be based on a weight measurement. However, volumetric dispensers are also used for powdered polyolefins and are readily adaptable for measuring vinyl plastisols in rotational casting.

Equipment through which inert gas and cooling media can be introduced into and vented from the mold during the heating and cooling cycles is recommended for rotational molding. It is shown schematially in Fig. 24. A system of this type will prove particularly useful in applications where comparatively high heating temperatures are necessary, such as when rotationally molding

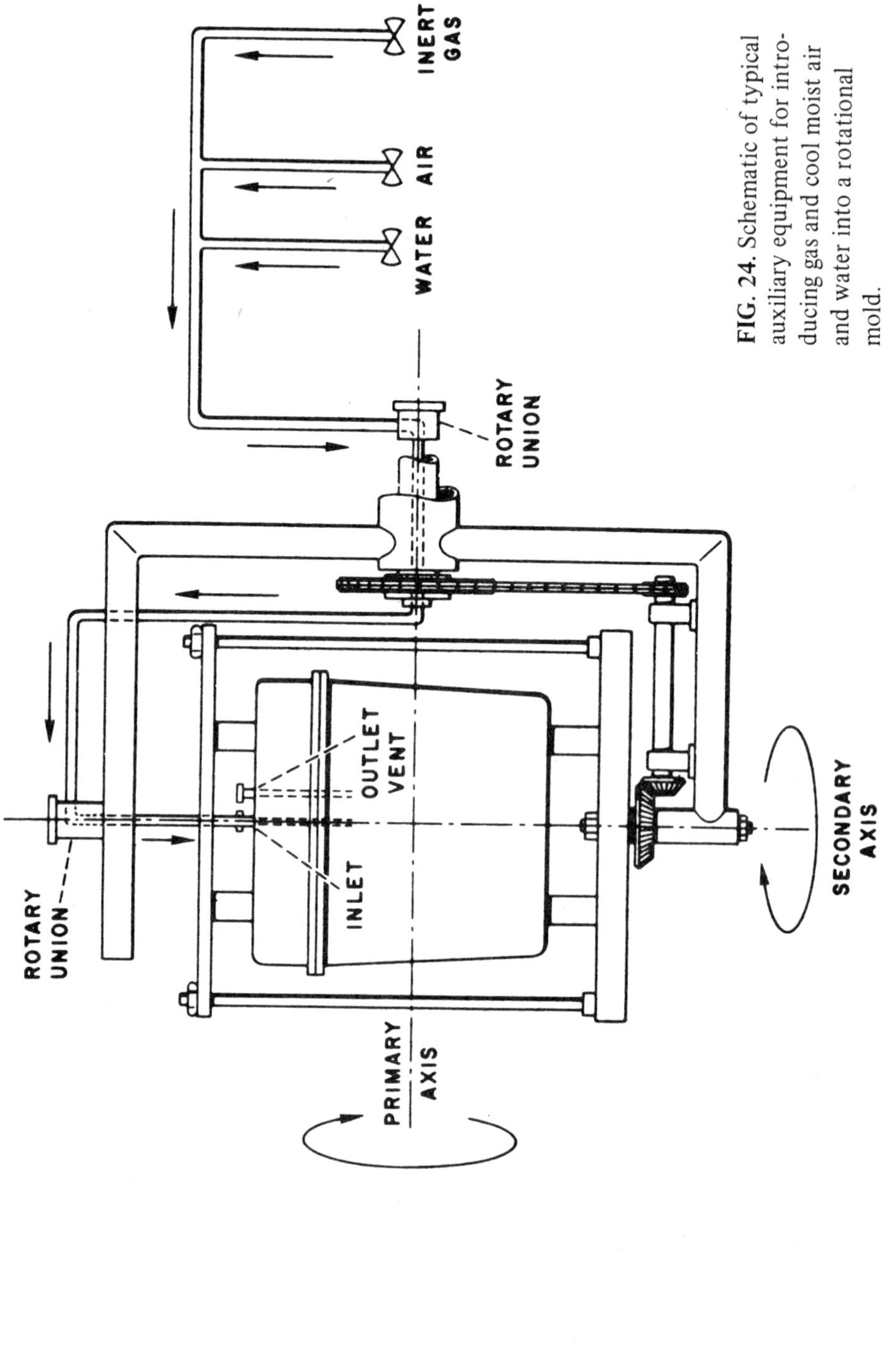

FIG. 24. Schematic of typical auxiliary equipment for introducing gas and cool moist air and water into a rotational mold.

high density (linear) polyethylene. It serves the purposes of (a) improved end product properties, (b) reduced odor and decoloration of the molded item, and (c) shortened cooling time.

Molding takes place in the presence of air with its oxygen content entrapped in the mold cavity. The introduction of an inert gas such as nitrogen or carbon dioxide into the mold during the heating cycle flushes out the oxygen and thus prevents thermal or oxidative degradation of the resin while it is being fused into the molded item. Such resin deterioration may occur if the resin is excessively heated. Thus, flushing out of the air will eliminate undesirable odor and discoloration in the interior of the molding, while simultaneously improving its physical properties.

Once the molded piece enters the cooling cycle, cool air can be substituted for the inert gas. This can be followed by the introduction of moist air and finally water to help cool the piece from the inside. Faster and more uniform cooling made possible by this technique (a) reduces cycling time, (b) decreases the crystalline content in the molded piece and thus improves its toughness, and (c) controls warpage as a result of simultaneously cooling the hollow molding from both its interior and exterior.

The inert gas and cooling media are introduced into the mold through hollow rotating drive shafts, mechanical seals, and a flexible high temperature hose inserted in the mold (see Fig. 24). To make the most efficient use of such auxiliary mold-flushing equipment, the following steps are observed:

(1) Enough inert gas is introduced into the mold to flush out all of the oxygen. Then the gas volume is reduced just enough to keep oxygen from reentering through the outlet vent.

(2) Shortly after the cooling cycle begins the gas is shut off and air is introduced into the mold.

(3) Once the interior layer of the item is cooled – the time required will vary with the size and shape of the moldings but should be ended shortly after the air is turned on – a small amount of water can be added to speed cooling. Then water input can be slowly increased until the piece is sufficiently cooled for removal from the mold.

When studying the economics of rotational molding, the most logical approach is to compare it to other molding techniques with which it is competing, namely, blow and injection molding. However, when comparing the cost per piece (unit cost) of the three molding processes, it must be kept in mind that these techniques have been designed for different purposes. Injection molding yields high-precision items at high production rates but cannot generally be used to make one-piece closed vessels. Blow molding is best suited for making bottles and similarly shaped items which require low melt index polyolefin resins. Rotational molding is best (though by far not exclusively) suited for producing closed, thick-walled items.

The low cost of rotational molds compared with blow and, especially, injection molds is estimated to amount to about 90% savings.

FIG. 25. Four-foot high mannequin, rotationally molded powdered polyethylene. Photo courtesy Hennessey Products, Ltd., Havant, England.

Changes in resin melt index and density affect the cycle time in rotational molding more than in injection and blow molding. However, increases in wall thickness affect the cycle time much less than in injection or blow molding. The wall thickness of a rotationally molded item can be increased by merely dumping additional powder into the mold; wall thickness increase in injection molding, on the other hand, requires a complete modification of the mold. Also, design alterations are usually made much more easily and at much less cost in rotational molding.

The marketing outlook for rotational molding indicates that the once-thought-of specialty process is emerging as one of the more versatile approaches to the manufacturing of an extremely broad range of products. The technique has already been accepted as particularly suitable for applications such as display window mannequins (Fig. 25), sporting equipment, and complex-shaped toys (Fig. 26) and other items for children.

FIG. 26. Hobbyhorse rotationally molded from powdered polyethylene. Photo courtesy Moulded Products, Inc., Maple Plain, Minn.

FIG. 27. Various sizes of traffic cones rotationally molded so that the square base can be opened by a flexible lip and filled with sand, gravel, or even water to prevent toppling over by wind or passing vehicles. Photo courtesy Heller Industries, Inc., Port Washington, Wis.

More recently, rotational molders have focused their endeavors toward products of the industrial type. Among the products now being evaluated are battery cases, traffic cones (Fig. 27), luggage, housings of various kinds (Fig. 28), waste receptacles, water and fuel tanks, and automotive parts including air ducts (Fig. 29) and fender liners.

OTHER POWDER MOLDING TECHNIQUES

There are three other techniques besides rotational molding known to be used to mold polyethylene powder. They account, however, for very little powdered polyethylene use.

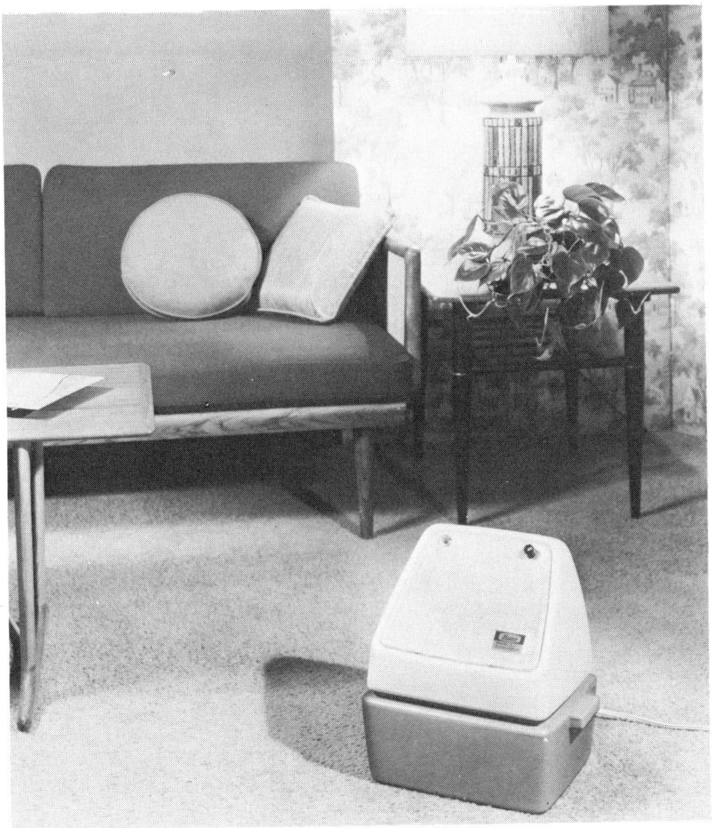

FIG. 28. Rotationally molded two-toned air-humidifier housing. Both housing sections were simultaneously molded with different colors. Photo courtesy W. B. Marvin Mfg. Co., Urbana, Ohio.

Similar in some respects to rotational molding is the so-called "Rock-and-Roll" method (U.S. Patent 2,881,476). In this process the mold is rotated only on the horizontal axis while the mold ends are rocked up and down. The double motion allows the powder to adhere to all interior mold surfaces during the heating period. Rotating of the mold (or molds) and seesawing of the whole frame is continued through the heating and cooling cycles.

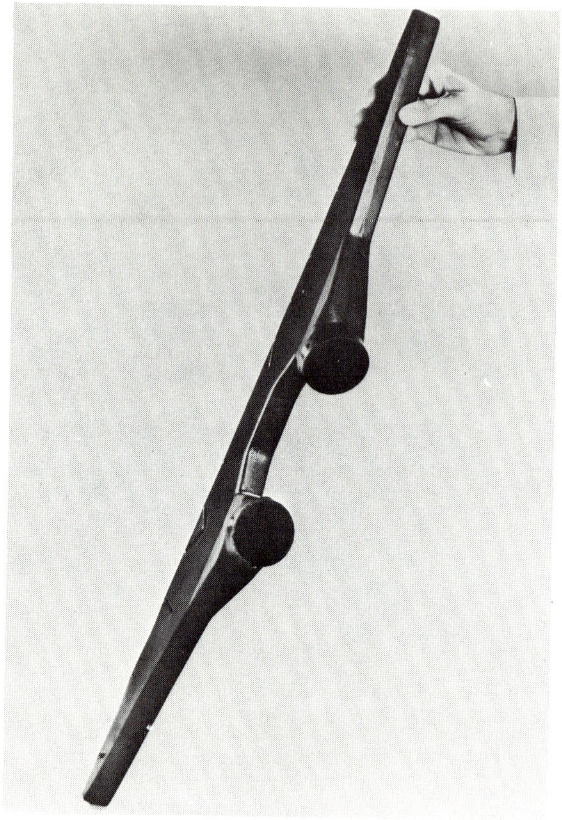

FIG. 29. Automotive air ducts. Photo courtesy Wiremold Company, Hartford, Connecticut.

The Engel process (U.S. Patent 2,915,788) uses a "static" method, i.e., the mold is not rotated, to produce molded parts. The cold mold is filled with powder, closed with a heat insulating cover, and placed in an oven operating at 500°-750°F (260°-400°C). The resin close to the inner mold wall melts and fuses into a continuous layer. The thickness of the fused layer is determined by oven temperature, heating time, and the melt index and density of the powder. Then the mold is removed from the oven, the excess or unfused powder is poured out, and the mold is returned to the oven to smooth the inner surface of the molded part. Finally, the mold is removed from the oven and cooled by spraying or immersion in a water bath.

The Heisler process (U.S. Patent 2,736,925) differs from both the previously mentioned methods of molding. In this process the mold is preheated and partially filled with powder. Molds with one closed end are filled in the upright position and then rotated about the long axis at a slight angle from the horizontal. Molds with both ends open are filled with an excess of powder and then seesawed. The fusing, removing of excess material, smoothing and cooling steps are similar to those of the Engel process.

REFERENCES

(1) E. S. Bobee, *Adhesives Age,* **2**, 30-37 (1959).
(2) F. J. Bockhoff, E. T. McDonel, and J. E. Rutzler, *Ind. Eng. Chem.,* **50**, 904-907 (1958).
(3) C. A. Burt, *Corrosion,* **16**, 26-27 (1960).
(4) R. L. Checkel, *Mod. Plastics,* **36**(2), 125-132 (1958).
(5) G. A. Curson, *Corrosion Prevent. Control,* (1956).
(6) D. R. Davis, *Plastics Technol.* **8**(6), 37 (1962).
(7) R. J. Fabian, *Mater. Design Eng. Spec. Rep.,* Vol. 38.
(8) J. Gaynor, *Chem. Eng. Progr.* **56**, 75-78 (1960).
(9) E. Gemmer, U.S. Pat. 2,844,489 (July 22, 1958).
(10) J. S. Heisler, A. Heisler, and A. J. Starr, U.S. Pat. 2,737,461 (March 6, 1956).
(11) S. E. Hmiel, U.S. Pat. 2,898,233 (Aug. 4, 1959).
(12) S. G. Howell and D. M. Akers, *SPE J.,* **21**, No. 6, 583-586 (1965).
(13) W. C. Johnson, *Symposium Rotational Molding, U.S. Industrial Chemical Co., Chicago, 1963.*
(14) E. P. Kieffer, U.S. Pat. 2,688,563 (Sept. 7, 1954).
(15) F. J. Nagel and D. C. Rickart, Polymer Corp. Report, Reading, Pa.
(16) J. A. Neumann, *Plastics Eng.,* (1950).
(17) *Mod. Plastics,* **43**(8), (1966).
(18) W. R. Pascoe, *Mater. Design Eng.,* **91** (Feb. 1960).
(19) J. Pinsky, A. R. Nielsen, and J. H. Parliman, *Mod. Packaging,* **28**, 145 (1954).
(20) Rheem Mfg. Co., *Chemical Week,* **92**(3), 35-37 (1963).
(21) "Rotational Molding of Microthene Polyethylene Powders," Booklet No. PL 23-865, U.S. Industrial Chemical Co.
(22) H. J. Schonhorn, *Appl. Polymer Sci.,* **8**, 355-361 (1964).

(23) R. B. Seymour, *Ind. Chem. Eng.,* **55**(9), 51-59 (1963).
(24) R. B. Seymour, *Plastics World,* **22**(4), 50-52 (1964).
(25) R. B. Seymour and F. Smith, *Ind. Chem. Eng.*, **56**, (8), 43-48 (1964).
(26) L. H. Sharpe, H. Schonhorn, and C. J. Lunch, *Intern. Sci. Technol.* **3**(28), 26-37 (1964).
(27) D. Taylor and J. E. Rutzler, *Ind. Eng. Chem.,* **80**, 928 (1958).
(28) L. S. Van Delinder, *Corrosion,* **15**, 33-36 (1959).
(29) P. West, *Mater. Design Eng.,* **61**(1), (1965).
(30) A. B. Zimmerman and D. R. Ellis, *SPE J.,* **22**, 41-45 (1966).

J. MACROMOL. SCI. – REVS. POLYMER TECHNOL., D1(1), 97-112 (1971)

Plastics and Other Polymers in Building

IRVING SKEIST AND JERRY MIRON

Skeist Laboratories, Inc.
Livingston, New Jersey 07039

ABSTRACT

Construction is the outlet for one-fourth of all plastics materials. On a tonnage basis, however, less than 2% of all building materials are plastics and other synthetic polymers (for adhesives, coatings, etc.). In a five year period during which tonnage of all building materials increased by less than 15%, plastics nearly doubled in usage. Their growth over the next five years will be somewhat slower, but still impressive, as they continue to displace older materials in various applications.

Plastics and related materials can offer esthetic appeal, light transmission, unlimited design and texture possibilities, resistance to water and corrosion, excellent dielectric characteristics, insulation against cold or heat, adhesion to other building materials, light weight, and high strength-to-weight ratio.

U.S. Dept. of Housing and Urban Development (HUD) is sponsoring more than a score of new housing techniques involving prefabrication of panels and modules. Some of these projects employ plastic techniques quite ingeniously, utilizing sandwich constructions, preassembled plumbing, drain, waste, and vent, and bathrooms, and even room-sized modules, filament wound.

Among the most rapidly growing uses for polymers in construction are polyvinyl chloride siding, fiberglass-reinforced plastic panels, polyurethane and polystyrene foam insulation. In the future, the high specific modulus that is available from graphite-reinforced fibers (many times as strong and rigid as steel on a volume basis) will foster the use of even these exotic materials in high-rise buildings.

INTRODUCTION

The earliest habitats of man, after he ventured out of the cave, were fashioned from inorganic plastics and natural organic polymers. Clay is a thermoset which hardens in the sun. The igloo is an architecturally correct dome made of blocks of thermoplastic, solid at ambient temperature, minimal in cost, easily cemented, and providing excellent thermal insulation. Wood is a composite, in which cellulose fibers reinforce the lignin binder. The thatched roof that one sees occasionally in the Dutch and Flemish countryside makes ingenious use of a natural insulating material.

The completely synthetic plastics, products of the twentieth century, have been used to a significant extent as building materials for only a few decades. Although some 25% of all plastics goes into construction, their volume to date is less than 2% of the total tonnage of building materials (1,2). However, the use of plastics in the United States nearly doubled during the five years ending 1969 (see Table 1), a period during which most other building materials increased in usage by only 10-15%.

This rapid rate of growth is expected to continue in view of the many advantages offered by plastics:

(1) High light transmission for glazing and lighting fixtures, combined with high impact strength, an especially important feature for schools and other public buildings.

(2) Colorability and esthetic appeal.

(3) Infinite texture possibilities or smooth surface if desired.

(4) Easy maintenance; permanence of color without painting.

(5) Unlimited design possibilities. They can be formed into arches, "hypars" (hyperbolic paraboloids), I beams, corrugated panels, ribbed structures, and other forms which increase strength and stiffness.

(6) Resistance to water, corrosion, weathering, making them suitable for bathrooms, plumbing, and rainwater goods.

(7) Excellent dielectric characteristics, hence, suitable for electrical insulation and conduit.

(8) Low thermal conductivity, consequently, warm and pleasant to the touch. Plastic foams provide exceptional insulation against cold or heat.

(9) Excellent adhesion (while still in liquid form) to other building materials.

(10) Light weight and high strength-to-weight ratio, an especially important feature in high-rise construction. Modules and prebuilt components can be

Table 1. Brighter Outlook for Construction Materials[a,b]

Year	New construction[c] Private	New construction[c] Public	Lumber, wood products, billion bd. ft	Plastics billion lb	Paint, varnish, lacquer, million gal	Portland cement, million bbl	Gypsum products, billion sq ft	Asphalt products, million squares	Aluminum products, billion lb	Copper products, billion lb
1964	$45,810	$20,390	30.0	2.5	394.5	366.3	9.3	105.7	1.7	1.5
1965	50,253	22,066	30.3	2.9	410.7	373.6	9.7	106.2	1.9	1.6
1966	51,120	24,000	29.2	3.4	416.3	380.7	8.4	99.8	2.0	1.6
1967	50,587	25,573	28.2	3.7	398.4	374.2	8.3	106.6	1.9	1.3
1968	56,996	27,696	30.7	4.4	423.7	397.3	9.4	108.4	2.3	1.3
1969	61,600	29,400	32.2	4.8	443.0	405.0	10.5	109.3	2.5	1.4

[a]From Ref. (2).

[b]Data (all dollar figures in millions) are based on reports from U.S. Department of Commerce; The Society of the Plastics Industry; National Paint, Varnish & Lacquer Assn.; Portland Cement Assn.; Gypsum Assn.; Aluminum Assn.; Copper Development Assn, and *Chemical Week* estimates for 1969.

[c]Private expenditures include single and multiunit residences and nonresidential buildings; public expenditures include residences, nonresidential buildings, utilities stations, railroads, and highways. Construction materials data indicate shipments, except paint, varnish, lacquer figures, which are for production-for-trade sales. Lumber, wood products figures are for softwood lumber for most part (hardwood flooring is included).

transported readily, permitting prefabricated construction, with great savings in labor. Construction can proceed throughout the year, and capital investment is tied up for shorter periods of time than with conventional on-site building techniques.

The U.S. Department of Housing and Urban Development (HUD) has sponsored a series of exploratory techniques in mass production of housing, now woefully lagging. The ingenious employment of plastics and composites in some of these developments will be discussed at length later in this paper (3).

There are also various disadvantages:

(1) Some of the physical properties are not as good as those of some other building materials. Scratch resistance, important in windows and bathroom fixtures, is less than that of glass or porcelain. Like wood and other organic materials, most plastics burn. (But self-extinguishing or nonburning grades of many plastics are now available.) Ultraviolet light causes deterioration, especially in combination with heat, often making plastics unsuitable for roofs and other outside surfaces unless pigmented or protected by uv inhibitors. Strength and modulus (stiffness) are not as high as those of some other materials; but on a volume basis (specific strength and specific modulus), plastics compare favorably, especially when reinforced, used in sandwich construction, or used in curved shapes not feasible with many conventional building materials.

(2) Cost per pound is relatively high. Again, the low density results in a cost per volume which makes plastics more comparable to other materials. Furthermore, the trend of plastic prices is generally downward, unlike other materials.

(3) Opposition from craft unions, especially plumbers and carpenters, and restrictive building codes have held back the use of plastics. Prefabrication is a partial answer to the former problem, while building codes based on performance rather than specific materials are becoming more widely accepted.

(4) The builder and architect seek more engineering and performance data before switching to these novel materials.

New construction in the United States amounted to $91 billion in 1969. Of this amount, $61.6 billion was for private and $29.4 billion for public construction. Less than one-third of the total expenditure went into private homes, the balance into industrial, commercial, institutional, and public works. Although the dollar value was 50% greater than that for 1964, the tonnage of materials used, as noted previously, increased by only 10-15%. Such ahs been the rampant surge of inflation in building costs.

Table 2. Plastics and Other Polymers in Construction (United States)

Material	Million pounds 1970	Million pounds 1975	Increase in 5 yr,%
Plastics, adhesives and foams			
Polyvinyl chloride	920	2200	
Phenol formaldehyde	600	900	
Urea and melamine formaldehyde	280	450	
Polyethylene	560	800	
Polystyrene	210	300	
Fiber glass-reinforced plastics	180	540	
Acrylic	160	270	
Polyurethane	120	300	
Miscellaneous	470	740	
Subtotal	3500	6500	86
Coatings			
Vegetable oil	350	300	
Alkyd resin	320	340	
Vinyl acetate polymers	225	340	
Acrylics	125	200	
Styrene butadiene	40	35	
Miscellaneous	50	65	
Subtotal	1110	1280	14
Carpeting, OEM	Small	50	
Sealants	30	40	
Total, rounded	4600	7900	71

The year 1970 will be a poor year for construction in the United States. High interest rates, inflation, and the economic downturn are all contributing factors which are expected to reduce this year's level of output. Nevertheless, plastics should show a significant gain through increased penetration.

Table 2 indicates our estimates of polymer usage in building for 1970 and 1975. Our figures are somewhat lower than those of others, principally because

Table 3. Polyvinyl Chloride in Building[a]

Building element	1968, million lb	1969, million lb
Flooring	330	?
Wire and cable	122	Steady
Siding	33	50-60
Storm windows and doors	11	12
Prime windows	7	13-15
Wall coverings	80	Good
Piping	72	105
Supply	(14)	(15)
Drain, waste, and vent	(16)	(30)
Gas	(3)	(6)
Sewer	(1)	(3)
Conduit	(20)	(35)
Fittings	(11)	(12)
Miscellaneous	(7)	(4)
Finishing and seals	25	Good
Shapes and profile	17	?
Other	18	?
Total	715	

[a]From Ref. (4).

we have adjusted the U.S. Tariff Commission figures downward for phenolic and amino resins used as adhesives and bonding agents. High pressure laminates, plywood, and particleboard all have some uses outside the construction field. In particular, much particleboard goes into furniture as well as for building applications such as underlayment, kitchens, and doors.

We have projected only a 70% increase in the use of polymers in building in the United States over the next five years, taking into account the likely continuation for some time of high mortgage interest rates and other factors which will slow down new construction. For plastics, adhesives and foams, however, an 85% increase is predicted over the next five years.

Polyvinyl chloride (PVC) is in every advanced country the dominant polymer used in building. We have assigned it 18% of present poundage, increasing to one-third of the volume by 1975, almost 2½ times its present volume, and in accord with its high growth rate until now. Low in cost, it excels as a dielectric in water and corrosion resistance, weathering (when properly formulated), resistance to indent and abrasion in resilient flooring.

An important factor in maintaining this rapid rate of advance will be the increased usage of rigid PVC and CPVC (chlorinated PVC) for siding, windows, and piping.

Mr. John Stahl of B.F. Goodrich Chemical Co. has provided a detailed breakdown of PVC usage in building (see Tables 3 and 4) (4).

Table 4. Rigid PVC[a] Growth in Building and Construction[b]

Year	mm lb
1964	160
1965	191
1966	228
1967	294
1968	365

[a]Excludes wire and cable and flooring.
[b]From Ref. (4).

A considerably greater usage for PVC in plastics pipe, tube, and fittings has been compiled by the Plastics Pipe Institute. No doubt these figures include chemical plant and other nonconstruction uses (see Table 5) (5).

In other countries, PVC has long been the predominant plastic material for construction, to a far greater extent than in the United States. For example,

Table 5. Sales of Plastics Pipe, Tube, and Fittings (United States)[a]

Material	1968, mm lb	1969, mm lb	Percent change
Polyethylene	80.	100.	25.0
Polyvinyl chloride	175.	225.	28.6
Acrylonitrile-butadiene-styrene polymer	51.5	60.	16.5
Styrene-rubber resin	58.	65.5	12.9
Other[b]	4.	6.	50.0
Fittings	57.5	68.	18.3
Total	426.	524.	23.0%

[a]From Ref. (5).

[b]Mostly polypropylene. The volume of CPVC has been almost insignificant until now.

in the United Kingdom, PVC constitutes approximately 60% of plastics (including foam but not adhesives and coatings); see Table 6 (6).

Table 6. Plastics usage in the United Kingdom[a]

Material	Thousand tons 1968	Thousand tons 1969
Polyvinyl chloride	67.1	92.8
LDPE (low-density polyethylene)	18.2	20.4
HDPE (high-density polyethylene)	1.5	2.1
Polystyrene	3.4	4.2
Polyester	6.9	9.0
Other	19.8	23.8
Total	116.9	152.3

[a]From Ref. (6).

An increase in PVC usage has been particularly impressive in gutters, downpipe, fittings, and soil pipe. An estimate by Mr. G. J. Sleddon of ICI, including adhesive usage for phenolic and amino resins, still assigns more than half the volume to PVC (7).

Table 7. Plastics Products for the Building Industry (Germany)[a]

		1967 increase		1968 increase
Value (million DM)	2,068	2.8%	2,218	7.3%
Weight (tons)	577,000	9.3%	661,000	14.6%
Estimated increase of capacity with fully synthetic fibers (tons)			34,400	20.5%
Total tons			695,033	

[a]From Ref. (9).

Table 8. Production Value in Building Industry (Germany)[a]

Production value (1000 DM)	1967	Increase, %	1968	Increase, %
Floor and wall coverings	632	0.4	602	−4.7
Foam plastics	268	3.4	269	0.1
Laminated sheets	238	3.	253	6.3
Pipes	202	5.2	249	23.3
Profiles	161	0.1	198	23.
Building components	128	16.3	152	18.7
Building plates, sheets	131	7.3	138	5.3
Revolving shutters	85	−2.3	88	3.5
Total[b]	2068	2.8	2218	7.3

[a]From Ref. (9).
[b]Total (including uses not listed).

In Germany, we see the same trend: a rapid increase in the use of pipe and profiles (8). The 5% drop in floor and wall coverings from 1967 to 1968 is largely because of replacement by carpeting (see Tables 7 and 8) (9).

Japan has long used PVC for both exterior and interior cladding (10). Corrugated PVC panels are employed for roofs despite the high uv exposure. Siding, flooring, pipe, and accordion-type doors are other important PVC products. Rigid PVC has gained good acceptance in pipe.

In the United States, polyethylene is used mainly for wire coating, but also as a vapor barrier for gas and cold-water piping. In Japan, several companies are producing polyethylene foam with densities typically as low as 2 lb/cu ft. Cross-linking, either by irradiation or preferably with the aid of a peroxide such as dicumyl peroxide, results in improved tensile strength and resistance to deformation. The foam is utilized as an insulation material, bonded to steel, aluminum, and other facing materials. Polypropylene is also foamed and utilized in the form of filament for tatami mats to replace the traditional rush, which has been increasing in cost because of reduced available acreage and poor harvests.

Polystyrene foam and rigid polyurethane foam compete with the traditional phenolic-bonded glass wool for insulation. At densities of slightly greater than 2 lb/cu ft, these materials are far more efficient than greater thicknesses of glass wool. The polystyrene is available in slabs, which must be adhesive-bonded to facings; the polyurethane can be foamed in place. This capability makes it convenient to use polyurethane in igloos, domes, and free-form structures. The departments of architecture at Yale, University of Michigan, MIT, and elsewhere have demonstrated the versatility of these materials.

The recent earthquake in Turkey made thousands of people homeless. A rescue operation made use of 16-ft diamurethane foam igloos, 10 ft high, capable of housing up to 10 people. The components of the foam were sprayed upon a dome-shaped PVC balloon fastened to a rotating disk. Construction time was only one hr. Density of the foam is 4 lb/cu ft.

The foaming machine, supplied by Bayer, A.G., together with the urethane components, is portable and is transported by truck to the building site. The finished polyurethane shell weighs only 440 lb and can be lifted from the disk by six strong men. Life of the igloo is predicted to be 10 years (11).

Polystyrene also is utilized for flooring and lighting; the latter application requires a nonflammable grade.

Seamless flooring, an innovation of the past several years, makes use of the abrasion resistance, transparency, easy maintenance, and resilience of solid polyurethane. A floor specialist expects a doubling of material requirements in the next five years. Decorative chips are sprinkled over a base coat of polyurethane or epoxy, then several layers of moisture-curable isocyanate prepolymer are laid down from solution.

Table 9. 1969 Production of all Plastics and FRP[a]

	Million pounds			Pounds per year per capita		
	All plastics	FRP	FRP in bldg.	All plastics	FRP	FRP in bldg.
United States	17,070	943	108	85	4.8	0.54
Japan	9,400	?	66	92	?	0.71
West Germany	8,020	120	38	138	2.0	0.64
France	2,400	90	33	47	1.8	0.66
United Kingdom	3,130	93	36	57	1.69	0.64
Italy	3,510	38	19	66	0.73	0.37

[a]From Ref. (12a-c) and authors' calculations.

Fiber glass-reinforced plastics (FRP) are one of the fastest growing materials in the building industry. Construction is the third largest market for reinforced plastics after boats and land transportation. Volume may triple in five years. Until now, the chief building uses for FRP have been in panels, often in the form of sandwiches with cores of foam or aluminum. They combine good wearing properties with high aesthetic appeal, especially in public buildings such as churches. Now, other areas that have been popular abroad are opening up in the United States: bathroom modules, tubs, and showers.

Although the United States lags somewhat behind Japan and considerably behind Germany in per capita production of all plastics, we are by far the leading producer per capita of reinforced plastics, as shown in Table 9. However, several countries exceed us in per capita usage of reinforced plastics for building. Once again we see the stultifying effect of the myriad United States' building codes, which the manufacturer is still required to satisfy, and the effect of the opposition from craft unions. In Japan, for example, bathtubs, bathroom units, and septic tanks already account for 40% of reinforced plastics (RP) usage (12a-c).

Acrylics are the glamour materials of the plastics industry. As lighting fixtures, glazing, skylights, and transparent or translucent panels, they are both attractive and functional. In the United Kingdom, vacuum-formed acrylic sheet in pastel colors has been employed for decorative bathtubs and washbasins.

Many other plastics materials go into specialized building applications. Coumarone indene and petroleum resins are fighting a losing battle for the floor tile market. Epoxy resins, with their excellent adhesion to concrete, form a substrate for industrial and institutional seamless flooring as well as a matrix for terrazzo flooring. Cellulose esters and polycarbonates compete with polystyrene, acrylics, and PVC for lighting applications. Polyacetals and nylon are utilized for plumbing fixtures and other hardware.

RECENT DEVELOPMENTS

Operation Breakthrough

According to HUD (3), the minimum needs for housing in the United States will amount to 26 million units within the next 10 years. Comparing this number with the present production of less than 1.5 million units per year (valued at $20 million), one must conclude that the tempo of construction must be at least doubled to fulfill this aim. As stated before, higher costs for conventional housing, increasing mortgage rates, and inflation are adding to the urgency of the problem.

A solution seems to be the initiation of massive projects for construction of low-cost standardized units. This, in turn, requires new methods of construction based mainly on prefabricated units or modules, cheap and easily processed materials, and also the overcoming of existing barriers of building codes and material standards based on conventional techniques of construction. As a by-product this effort could discharge much of the unrest accumulated within overcrowded inner cities and might also decrease the influx from rural areas.

The most significant program which has been initiated by the U.S. Government is HUD's "Operation Breakthrough." Under this plan, companies have submitted proposals for entire low-cost housing systems as well as for specific parts or aspects of housing; several hundred such plans have been offered. Of these, some twenty have been chosen and will be tested experimentally, at a Federal-private cost of $75 million. A total of $3 billion is to be expended during the next decade.

Current conventional housing prices have a median value of $17,000-20,000/unit, stemming from construction costs of $13/sq ft of living space. To provide shelter for an estimated 9 million people living presently in inadequate conditions, this cost will have to be reduced to less than $10/sq ft. This can be done only by promoting large industrialized building projects. At the present time, construction is predominantly in the hands of small builders with an average production of perhaps 100 houses per year. There are at least 50,000 home builders in the United States, none having more than 1% of the total market (13). It is very difficult to economize under these conditions.

The Materials Systems Southwest Corp. (Valley Center, Calif.) has utilized reinforced plastics to build low-cost, attractive homes for Indians at the Pauma, Rincon, and La Jolla reservations in San Diego County. A three-bedroom home costs $10,000. Reinforced plastics are used for beams, roof, and doors. The goal is 3 million new homes for Indians, Mexican-Americans, and blacks. A new state-wide code, superseding local codes, was necessary to permit the new construction (14).

Pemtom, Inc. (Bloomington, Minn.) has developed a factory-produced stressed-skin plywood module joined by means of a polymer bond into a single structural unit. These "Unimods" may be arranged and stacked to make multifamuly high-rise or single-family detached dwellings.

Ball Brothers Research Corp. (Boulder, Colo.) will utilize a foamed plastic core, structural panel system which can accommodate standardized doors, windows, and other hardware. Reactants that result in polyurethane foam are poured between sheets of hardboard and cement asbestos board. With the aid of epoxy resin as a matrix, the exterior is covered with a layer of stone or other aggregate.

Christiana Western Structures, Inc. (Los Angeles, Calif.) has developed a shop-fabricated wood frame panel for walls, partitions, and roof construction. A special polyester resin finish coat, 1/32 in. thick, reinforced with fibrous glass, is applied to a full length plywood wall base, permitting the fabrication of a complete wall panel of any desired length without a joint.

TRW Systems Corp. (Redondo Beach, Calif.) has developed the most innovative of the systems. First, panels and small shapes will be produced in an off-site factory. A typical sandwich panel may be built up from a paper honeycomb core (bonded with phenolic resin) between layers of polyester reinforced with woven roving, gypsum wallboard, and facings of polyester reinforced with chopped fiber. At the building site, a relocatable factory will produce room-sized modules on a mandrel which is capable of assuming many different configurations. During the wrapping process, services will be incorporated into the module under factory conditions.

RP for High-Rise Buildings

The low-cost housing possibilities of reinforced plastics are by no means limited to one- to two-family houses. Experimental high-rise 22-story buildings are being constructed in London. These 200-apartment units have outside walls made of fiber glass skin panels with concrete foam cores and are flame resistant.

Preassembled Plumbing

Polyvinyl chloride, acrylonitrile-butadiene styrene (ABS), and CPVC pipe are being used ever more extensively wherever building codes permit for drain, waste, and vent (DWV) application, as well as for water distribution. The cost savings are significant – $40 to $90 per housing unit.

Economies are realized in both labor and materials. Factory labor is approximately $3.50/hr, compared with $7-10/hr for union plumbers. Prefabricated DWV systems use far less material than those assembled at the job site because of the waste inherent in custom cutting of pipe. Also, with trained labor, prebuilt units require 20% fewer man-hours for both factory and on-site operations than conventional on-site installations. Finally, because factory operations are independent of the weather, construction time is reduced.

Modular bathrooms are another fast-growing outlet. They are more resistant to corrosion and lower in cost than conventional units. They are usually made from glass fiber-reinforced polyester by the spray-up technique. Some 200,000 units were sold in 1968, and 50% more in 1969. By 1975, modular bathrooms can capture as much as 30-40% of the total market. One popular unitized bathroom costs approximately $1200, compared with $1600 for a conventional bathroom with cast iron tub.

Mobile Homes

While the construction of single-family conventional homes fluctuates with economic conditions, the number of mobile homes built has continued to grow each year for the past eight years, so that it now constitutes one-fourth of all housing units. Introduced initially as trailers, primarily for travel, they are now mainly stationary once transported from the factory. Cost last year averaged $8.30/sq ft as compared to $13/sq ft for conventional housing. Depending on size, these homes sold at $3,000 to $15,000.

Various plastics are used here for pipe and fittings; polyethylene film as vapor barrier, ABS and vinyl for DWV. Other plastics applications are water tanks, furniture, decorative laminates, plumbing and lighting fixtures, and wire insulation. In the future, foam insulation – 2400 sq ft/unit will probably replace glass wool, and RP panels or thermoformed ABS may be used instead of aluminum siding.

Carpeting as "OEM" (Original Equipment of Manufacture)

An important new development in low-cost one-family homes is the installation of wall-to-wall carpeting as part of the original structure. Since the underlayment may be particleboard or even concrete rather than a more expensive material such as oak flooring, the finished house is cheaper and at the same time more comfortable. The FHA has established standards for acceptable carpeting (15).

Graphite Fiber Reinforcement

In the future, it is entirely possible that graphite fiber will be used as a reinforcing and structural material for skyscrapers. Developed principally in the United Kingdom, Japan, and the United States, these materials, available with moduli upward of 50 million psi, are being sold in only small quantities in excess of $100/lb. At these prices, one can only afford to use them in aerospace and other specialized applications where weight savings is critically important. On a weight basis, these materials are many times as strong and rigid as steel. Within several years as usage increases, the price of graphite fiber will be reduced to the point where it can economically be substituted for steel in skyscrapers. In Japan, where the ever-present danger of earthquakes has required much more sturdy and costly buildings than elsewhere, the prospects for graphite-reinforced structural members is especially bright.

REFERENCES

(1) I. Skeist, ed., *Plastics in Building,* Reinhold New York, 1966.
(2) Anon., *Chemical Week*, 14 (Aug. 19, 1969).
(3) HUD News, U.S. Dept. of Housing and Urban Development, Washington, D.C., release of Feb. 26, 1970.
(4) J. Stahl, B.F. Goodrich Chemical Co., private communication; also, R. Holtz, B.F. Goodrich Chemical Co.
(5) Plastics Pipe Inst., Div. SPI, New York, N.Y.
(6) Anon., *Brit. Plastics,* 56 (Jan. 1970).

(7) G. J. Sleddon, *Brit. Plastics,* 78 (Jan. 1970).
(8) A. Schwabe, *Kunststoffe,* **59**, 723 (1969).
(9) Dipl. Ing. Amtor Schwabe, Darmstadt, West Germany, private communication.
(10) Anon., *Japan Plastics Age (Osaka),* 8(4), 25 (1970).
(11) Anon., *Product Eng.* 54 (May 11, 1970).
(12) (a) *Mod. Plastics,* (April 1970); *ibid.,* (Jan. 1970). (b) *Plastics World,* (Dec., 1969). (c) SPI RP/Composites Conf., Washington, D.C., Feb., 1970.
(13) L. A. Mayer, *Fortune,* (Dec. 1969).
(14) *New York Times,* Sec. 8, p. 1, Oct. 26, 1969.
(15) J. Kanegis, Federal Housing Administration, Dept. of Housing and Urban Development, Washington, D.C., private communication.

Fire Retardance of Polymeric Materials

IRVING N. EINHORN

University of Utah
Salt Lake City, Utah

ABSTRACT

The flammability characteristics of natural and synthetic polymeric materials is of broad interest with particular aspects including intumescent coatings, fibers and fabrics, reinforced plastics, elastomers, cellular plastics, and ablative materials. Rapid expansion of plastics in wearing apparel, draperies, carpeting, automotive and aircraft components, and in allied consumer industries, makes it imperative that a better understanding of the flammability characteristics of materials be gained.

THEORETICAL CONSIDERATIONS OF FACTORS RELATING TO COMBUSTION OF PLASTICS

The burning process of a unit mass of material can be considered to proceed in several stages, depending upon the source of ignition, geometry, attitude, environment, and material characteristics, both molecular and macroscopic The flow chart (Fig. 1) illustrates a schematic representation of the stages mentioned previously.

Stage I-A – Primary Thermal Processes

Heat from an external source is applied to the polymeric material causing a gradual rise in temperature. The rate of temperature rise depends on the temperature and flow rate of the heat source, the temperature differential, as well as the basic characteristics of the material such as specific heat, thermal con-

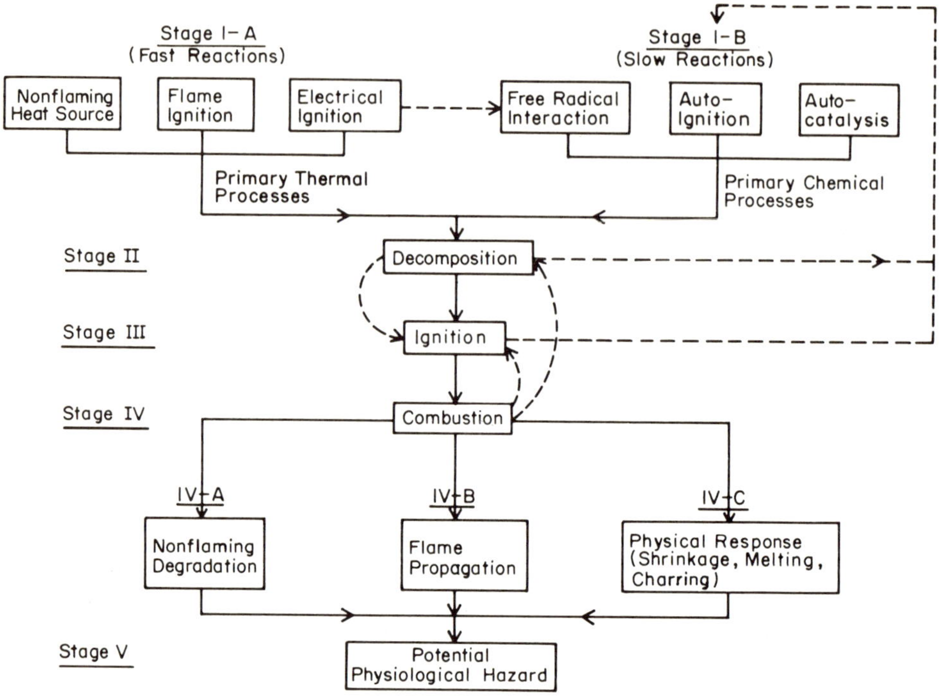

FIG. 1. Schematic representation of combustion and thermal degradation.

ductivity, heat of chemical reaction, heat of fusion, heat of vaporization, and other changes that may take place during the heating of the material.

Stage I-B – Primary Chemical Processes

In addition to thermal energy, the external heat source may also introduce free-radical species for chain branching reactions. Also, the heated material can be autoreactive because of either a high heat of reaction or the generation of its own catalyst. In either case, the reaction rate is further accelerated in a manner which can be independent of the primary thermal process. Lastly, the heated material can generate gaseous or condensed species which can react with gaseous or adsorbed oxygen. In general, the accumulation of energy by such bootstrapping chemical processes is initially slow.

Stage II – Decomposition

As soon as the materials reach their decomposition temperature, the following types of products may be formed as seen in the schematic representation of polymer degradation (Fig. 2). Elimination of combustible gases, i.e., gases

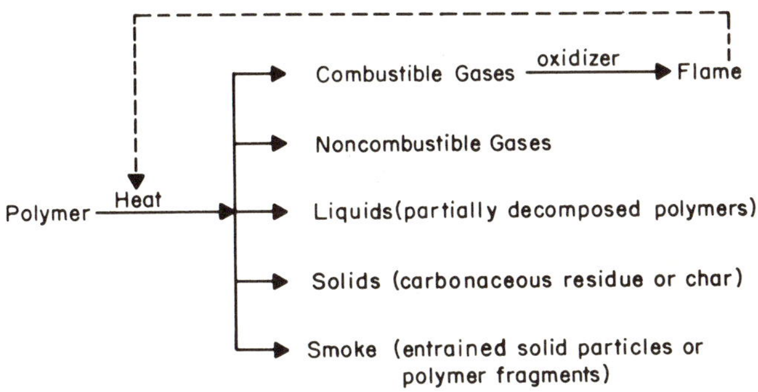

FIG. 2. Schematic representation of polymer degradation.

that will burn in the presence of air, would effectively preclude burning. However, in most instances, this is impossible since most of the organic material cannot be reduced to a highly carbonaceous residue without the release of some volatile hydrogen-containing compounds. Typical combustible gases are hydrogen, methane, ethane, ethylene, formaldehyde, and carbon monoxide. Boettner [1] reported on the combustion products of several polyvinyl chloride polymers which were heated from 200° to 600°C at a rate of 3°C/min. The rate of air supply was 60 cc/min which, when integrated over the entire run, would result in about twice the amount of oxygen required to convert all of the carbon to carbon dioxide. However, this was not true for any particular time in the combustion process, in that, first, the carbon was not available to form carbon dioxide until higher temperatures were reached, and, second, the air flow pattern was such that all of the oxygen was not in the immediate vicinity of the sample. Table 1 shows the volatile products of combustion identified in Boettner's studies.

Typical noncombustible gases (i.e., gases that do not burn in the presence of air) include hydrogen chloride, hydrogen bromide, carbon dioxide, and water vapor. The evolution of noncombustible gases which may be formed during combustion tends to reduce burning by several mechanisms: (1) The

Table 1. Volatile Combustion Products of Polyvinyl Chloride Polymers

Compound	Polyvinyl chloride homopolymer (MW 109,000)	Copolymer of vinyl chloride and vinyl acetate (85:15) (MW 55,000)
HCl	583.00	500.00
Acetic acid	–	0.60
CO_2	729.00	923.00
CO	442.00	292.00
Methane	4.60	4.40
Ethylene	0.58	0.60
Ethane	2.20	2.30
Propylene	0.47	0.56
Propane	0.84	0.88
Vinyl chloride	0.60	0.72
1-Butene	0.18	0.22
Butane	0.28	0.29
Isopentane	0.02	0.02
1-Pentene	0.06	0.09
Pentane	0.16	0.21
Cyclopentene	0.05	0.05
Cyclopentane	0.05	0.06
1-Hexene	0.05	0.08
Hexane	0.12	0.17
Methylcyclopentane	0.04	0.05
Benzene	36.00	28.00
Toluene	1.30	0.96

heavy gases may shield the oxidizer from the burning mass, (2) all noncombustible gases help dissipate the heat and cool the fire, and (3) the latent heat of vaporization of water tends to reduce the temperature. The most desirable products of decomposition are the solid carbonaceous residues or chars since they help to preserve structural integrity of the material, retard the outward diffusion of combustible gases, and thermally insulate adjoining masses from temperatures which produce thermal degradation. Char structures also prevent mixing of air with combustible gases. The decomposition stage is greatly affected by the following factors:

1. The temperature of initial decomposition.
2. The heat of decomposition, or heat absorbed or released during decomposition.
3. The manner in which polymers decompose, i.e., the relative amounts of the various products of decomposition.

DiPietro [2] classified polymers into two categories: those which decompose to yield mostly monomer (Table 2), and those which decompose to yield little or no monomer (Table 3).

Table 2. Thermal Decomposition of Polymers which Form Mostly Monomer

Polymer	Temp of pyrolysis, °C	Conversion to monomer, %
Polymethacrylonitrile	250	85
Polymethyl methacrylate	350	100
Polystyrene	350	65
Poly-α-methyl styrene	350	100
Polyisoprene	370	46
Poly-m-methyl styrene	390	44
Polyisobutylene	400	46
Polytetrafluoroethylene	600	86

Stage III – Ignition

In the presence of a sufficient amount of oxygen or an oxidizing agent the combustible gases ignite. Ignition depends on the following characteristics of the material:

1. The flash ignition temperature, i.e., the temperature at which gases evolved from the material can be ignited by a flame or a spark.
2. The autoignition temperature, the temperature at which the reactions within the material become self-sustaining to the point of ignition.

Table 3. Thermal Decomposition of Polymers which Form Little or No Monomer

Polymer	Complete pyrolysis, °C	Decomposition products		
		Low boiling	High boiling	Residue
Polyvinyl acetate	280 (vacuum)	Acetic acid hydrocarbons		Carbon
Polyacrylonitrile	350 (N_2)	NH_3, HCN	N_2 containing 66% C fractions	
Polypropylene	410 (vacuum)	Hydrocarbons	Paraffin fraction 97% MW (200)	
Polyvinyl chloride	440 (N_2)	51.9% HCl, unsaturated hydrocarbons	Unsaturated hydrocarbons	23% with 1.2% Cl_2
Polyvinylidene chloride	440 (N_2)	67.5% HCl		28% with 12.1% Cl_2
Polybutadiene	420 (vacuum)	Hydrocarbons	93% fractions MW 666-784	Little
Polyethylene	475 (vacuum)	Hydrocarbons C_3—C_{15}	Paraffin fractions-96% MW 690	

3. The minimal level of oxygen necessary to sustain ignition and combustion.

Seader [3] illustrated the effect of autocatalysis by comparing the thermal degradation of polytetrafluoroethylene and polystyrene. Figure 3 indicates

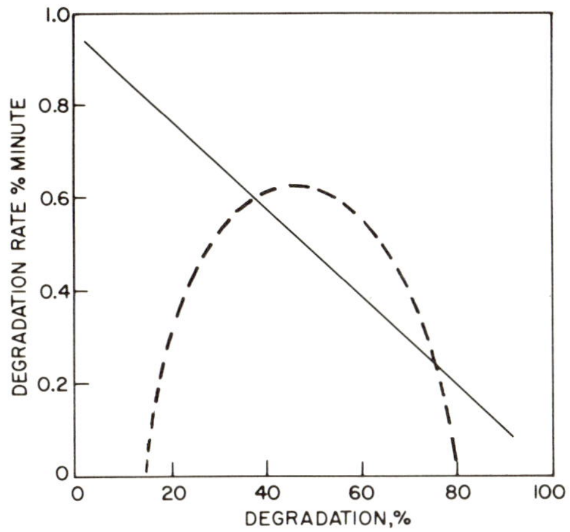

FIG. 3. Degradation of polytetrafluoroethylene: (——) PTFE and (- - -) PS.

that the polytetrafluoroethylene exhibits a linear relationship between degradation rate and percent decomposition when studied under isothermal conditions. The principal product of degradation is $\left(\begin{matrix} F & F \\ C = & C \\ F & F \end{matrix}\right)_n$. Polystyrene degrades as though the process were autocatalytic. At a given temperature, the experimental degradation rate first increases to a maximum rate and then decreases. Detailed experimental data on the thermal degradation of polystyrene (MW 230,000) are in reasonable agreement with the emperical autocatalytic rate law:

$$\frac{-dw}{dt} = A \exp(-E/RT)(\alpha w + \beta w^2 + \gamma w^3) \qquad (1)$$

where w is the instantaneous weight of the polymer residue material during the degradation process, t is the time, A is the pre-exponential factor, E is the activation energy, R is the universal gas constant, and T is the absolute temperature.

Fenimore and Martin [4] designed a test procedure to measure the limiting oxygen concentration or the minimum level of oxygen required to sustain ignition and combustion. For normal use, a material can be considered self-extinguishing if it cannot continue burning with less than 21% oxygen and nonigniting if it cannot be ignited with less than 21% oxygen. Table 4 presents the limiting oxygen concentrations for selected polymeric materials.

Table 4. Limiting Oxygen Concentrations for Selected Polymeric Materials

Polymer	$n\,O_2/(n\,O_2 + nN_2)$
Polyoxymethylene copolymer	0.148-0.149
Polyoxymethylene homopolymer	0.150
Polyethylene oxide	0.150
Polymethyl methacrylate	0.173
Polyethylene	0.174-0.175
Polypropylene	0.174
Polystyrene	0.181
Epoxide (conventional resin)	0.198
Epoxide (cycloaliphatic)	0.198
Chlorinated polyethylene	0.211
Polyvinyl alcohol	0.225
Polyvinyl fluoride	0.226
Chlorinated polyether	0.232
Polycarbonate	0.260-0.280
Polyphenylene oxide	0.280-0.290
Polyamide	0.290
Polyvinylidene fluoride	0.437
Polyvinyl chloride	0.490
Polyvinylidene chloride	0.600
Polytetrafluoroethylene	0.950

Stage IV – Combustion

The characteristics of the polymer in this stage may be correlated with energy relationships in the molecule. The net thermal effect of combustion can be either negative or positive. If it is negative, an external supply of heat is necessary to support combustion; if it is positive, the excess of heat is made available to increase the temperature of the adjacent material.

The most important energy factors are cohesive energy, hydrogen bonding, heat of combustion, and dissociation energy.

Cohesive Energy. The cohesive energy is concerned with secondary bond forces which lead to the aggregation of molecules in solid and liquid phases. It is the total energy necessary to remove a molecule from the liquid or solid material. Low cohesive energies permit the molecules to volatilize easily and thus aid in thermal decomposition. A molecule does not volatilize until the cohesive energy is exceeded. A direct relationship exists between melting point and cohesive energy. Compounds possessing low cohesive energies tend to melt and drip increasing flame propagation. Dickert and Toone [5] showed that cohesive energy increases with molecular weight and may be further increased by the introduction of polar groups into the polymer structure. If the cohesive energy exceeds the primary bond energies, the polymer molecules will decompose before they volatilize. This decomposition may lead to char formation. Cohesive energy values for chemical groups found in many polymers are given in Table 5 [5].

Table 5. Molar Cohesive Energy of Organic Groups

Group	Cohesive energy, kcal/mole
—CH_2— (Hydrocarbon)	0.68
—O— (Ether)	1.00
—COO— (Ester)	2.90
—C_6H_4— (Aromatic)	3.90
—CONH— (Amide)	8.50
—OCONH— (Urethane)	8.74
—NHCONH— (Urea)	10.64

A review of Table 5 indicates that the incorporation of ester, aromatic, amide, urethane, and urea groups into the polymer will raise the cohesive energy and would be expected to improve the flame resistance of the material.

Hydrogen Bonding. The secondary bonding between hydrogen and strongly electronegative atoms such as oxygen has a tendency to raise the cohesive energy and the melting point of the polymer. Therefore, in nylon and urethane polymers an increase in hydrogen bonding leading to higher melting points will tend to favor flame retardancy.

Heat of Combustion. The heat of combustion of the polymer, i.e., the energy that is relased during burning, is an important factor which affects the flame-resistant characteristics of a plastic. The heat liberated during burning of a polymeric material must be in excess of the ignition point or the flame will extinguish itself. Table 6 [5] lists the heats of combustion for various linkages found in polymers. Thus, as the concentrations of C═O or C—O bonds are increased in the polymer, the heat of combustion is reduced.

Table 6. Heats of Combustion of Chemical Bonds

Type of bond	Heat evolved, cal
C≡C	203,200
C═C	121,800
C—C	52,900
C—H	52,700
C—O	15,000
C═O	Zero

Dissociation or Bond Energies. The energy required to rupture bonds also is a factor controlling combustion of polymeric materials. The bond energies of linkages commonly encountered in plastics are presented in Table 7 [5].

A review of the data shown in Table 7 clearly indicates the desirability of building functional groups containing —C═O groups such as urethanes, ureas, amides, and esters to impart high dissociation energies to the polymer. It is necessary, on the other hand, to have some groups such as —C—Cl and —C—Br bonds which possess relatively low dissociation energies. Scission of these weak linkages results in the liberation of halogen atoms which are effective fire retardants.

Substage IV-A – Nonflaming Degradation

Many materials will degrade or smolder by a nonflaming propagation. In fact, some materials do not combust in normal oxygen environments. This is not to say that these materials are not hazardous for in the case of a fabric, if they transport heat at a rapid rate, they may cause severe local burns. With the exception of these materials with high thermal conductivity, the basic characteristics mentioned here are those desirable for wearing apparel.

Table 7. Dissociation Energies of Chemical Bonds

Bonds	Dissociation energy, kcal/mole
C—N	49-60
C—C	59-70
C—O	70-75
N—H	84-97
C—H	87-94
O—H	101-110
C═C	100-125
C═O	142-166
C—Br	54
C—Cl	67

Substage IV-B – Flame Propagation

Propagation results when the net heat of combustion is sufficient to bring the adjacent mass to the combustion stage. The ease with which the material is brought to this stage depends upon its chemical structure, geometry, and attitude. It should be apparent that an oxidizing agent is generally required to sustain combustion. Some materials, in their degradation process, release oxygen and then will support combustion at a limiting oxygen index of less than 21%; however, other materials require oxygen from the environment to support combustion. A thermoplastic polymer may self-extinguish after ignition occurs because the flame front melts away. The same thermoplastic polymer may undergo total combustion if reinforced with glass or asbestos fibers which permit propagation of the flame front by a wicking action.

The effect of weave, denier, and fiber composition will greatly influence flame propagation in fabrics. A typical example of the effects of fabric geometry can be seen in the comparison of the flame propagation characteristics of a loose knit sweater versus that of an angora-type fiber sweater. The loose weave fabric will burn with a slow propagation rate because of small precursor heating of adjacent material, while the angora fabric, with its high surface area, will enhance combustion and thus burn with a more rapid propagation. For this reason, propagation is often treated as a surface phenomenon and hence surface flame spread is considered as a realistic measure of propagation of materials.

Substage IV-C – Physical Response (Shrinkage, Melting, and Char Formation)

Straehl [6] noted that synthetic fabrics such as polyamides and polyesters shrink considerably at relatively low temperatures. This shrinkage can be serious in clothing fires. Synthetic underwear shrinks onto the body when the outerwear burns; it makes, furthermore, excellent thermal contact with the skin and if the underwear catches fire, the results are likely to be fatal.

The burning droplets from melting synthetics can make particularly deep and serious burns according to Crikelair [7]. Wool, as well as synthetics, can shed burning pieces, but the drips from wool can safely be caught by the hand because of the low thermal conductivity of the wool. The char from wool is a foam of low relative density, and like all foams is a good insulator and makes poor thermal contact with the skin. But the drops of many synthetics are molten and of moderate thermal conductivity – they stick and burn.

Einhorn and Mickelson [8] reported on the formation of char structure during the combustion of model urethane foams retarded with reactive and nonreactive fire retardants. The rigid urethane foam specimens, possessing a high degree of aromaticity in the polymer backgone and a cross-link density of less than 340, developed strong continuous char structure during burning. When the cross-link density exceeded 400, intumescence and slight melting occurred. As the cross-link density exceeded 500, there was no evidence of char formation and the samples were totally consumed. As the degree of aromaticity was reduced, the effective formation of char structure diminished. All flexible urethane foam samples, similar to that used in clothing insulation, burned to complete destruction. Evaluation of the data obtained in numerous laboratory experiments strongly indicated that the parameters that effect the development of improved thermal stability and flammability characteristics in cross-linked polymers are:

1. The degree of aromaticity in the polymer backbone
2. The nature and functionality of the monomers
3. The molecular weight per cross-link
4. The nature and method of incorporation of additives into the polymer system to retard combustion.

Stage V – Potential Physiological Hazard

Numerous test methods have been promulgated to evaluate the flammability characteristics of polymers. These methods of evaluating the combustion properties of materials may be divided into five general classes according to the properties they are intended to measure:

1. Ease of ignition
2. Combustion duration
3. Flame propagation
4. Fuel contribution
5. Gas and smoke evolution

Considerable variation may be encountered in the results obtained depending on the processing characteristics of the material, the molecular structure of the fabric, the method of fire retardation, and the variation in testing procedures. The methods used to rate the parameters of combustion do not take into consideration or attempt to evaluate the potential hazards to life of individuals involved in the fire.

THEORETICAL CONSIDERATIONS OF THERMAL DEGRADATION

Reaction Kinetics by Thermogravimetry

Thermogravimetry may be used to elucidate the reaction kinetics for solid or liquid-state thermal decompositions. In general, these reactions proceed stepwise through a series of intermediate reactions that involve concomitant formation and decomposition of several solids (s) or liquid (l) phases. Luckaszewski and Redfern [9] have summarized these solid-solid transformations as follows:

a. Reactions involving a decomposition such as:

$$\text{A (solid)} \longrightarrow \text{B (solid)} + \text{C (gas)}$$

b. Reactions between two solid phases:

$$\text{A (s)} + \text{B (s)} \longrightarrow \text{AB (s)} + \text{C (g)}$$

c. Reactions confined exclusively to the solid phase:

$$\text{A (s)} + \text{B (s)} \longrightarrow \text{A (s)} + \text{D (s)}$$

d. Reactions involving participation of liquid phase at some stage:

$$A\,(s) \longrightarrow B\,(s) + C\,(l)$$

e. Reactions involving a solid and a gas such as from a controlled atmosphere:

$$A\,(s) + B\,(g) \longrightarrow C\,(s)$$

The reactions illustrated above can occur consecutively or simultaneously. Thermogravimetry is generally used for reactions of single solids, hence (a) and (e) are most applicable. Differential thermal analysis (DTA) may be used to study reactions of the type (b), (c), and (d).

Two general methods are used in kinetics studies by thermogravimetry:

1. The nonisothermal method, in which a plot of weight change vs. temperature (and/or time) is obtained, under dynamic temperature conditions at a linear rate of heating.
2. The isothermal method, in which a plot of weight change vs. time is obtained, at constant temperature.

Nonisothermal Methods

Maximum Point Method. A quick approximation of the kinetics of thermal decomposition can be obtained from a single thermogram (Fig. 4) using the kinetic equation developed by Akita and Kase [10].

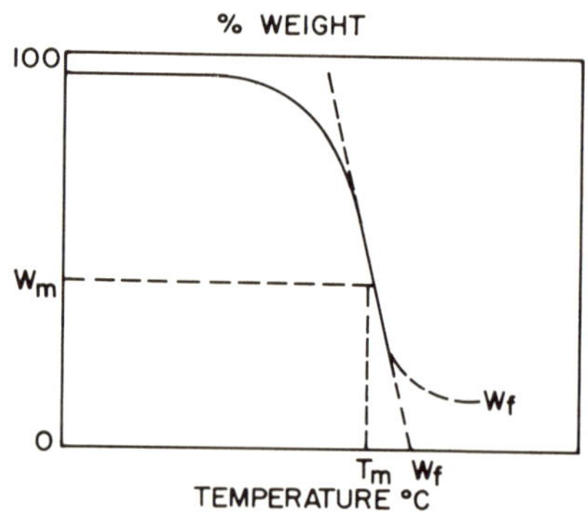

FIG. 4. Single thermogram kinetic plot.

$$\frac{d\gamma}{dt} = \frac{A}{b}\exp(-E/RT)\gamma^n$$

where

$$\gamma = \frac{W_m/W_0 - W_f/W_0}{1 - W_f/W_0} \tag{2}$$

Here, T – a + bt, E is the activation energy, n is the order of reaction, A is the frequency factor, R is the gas constant, W_0 is the initial weight; W_f is the final weight, W_m is the weight remaining at maximum rate of weight loss, and t is the temperature. The solution to the kinetic equation is given by

$$n = \gamma_m^{1-n} \tag{3}$$

$$E = \frac{nRT_m^2}{\gamma_m} \frac{-d\gamma}{dT_m} \tag{4}$$

$$A = \frac{bE}{RT_m{}^2} \exp(-E/RT_m) \tag{5}$$

where m corresponds to the maximum rate.

Multiheating Rate Methods. It is possible that a substantial error may occur if kinetic data are derived from a single decomposition thermogram. Flynn and Wall [11] have proposed a method for determining activation energies directly from weight loss vs. temperature data at several heating rates. They defined the degree of conversion, C, such that 1 – C, the residual fraction, equals the weight of material loss divided by the total weight loss at T or $t \rightarrow \infty$. The thermogravimetric rate is given by Eq. (6).

$$\frac{dC}{dT} = (A/\beta)f(C)\exp(-E/RT) \tag{6}$$

where T is the absolute temperature, β the constant heating rate, A the pre-exponential factor of the Arrhenius equation, E the energy of activation, R the gas constant, and f(C) the function of degree of conversion (weight loss).

Flynn and Wall [11] assumed that A, f(C), and E were independent of C. They then separated the variables and integrated to obtain

$$E \cong -4.35 \frac{d \log \beta}{d(1/T)} \tag{7}$$

Therefore, from weight loss vs. temperature thermograms at several heating rates, one may determine the corresponding temperatures at a constant weight loss. This was illustrated by Flynn and Wall (Fig. 5) for theoretical curves for a case often found in polymer decomposition where the isothermal rate goes through a maximum value. Then, from the slope of a plot of log β vs. 1/T, the

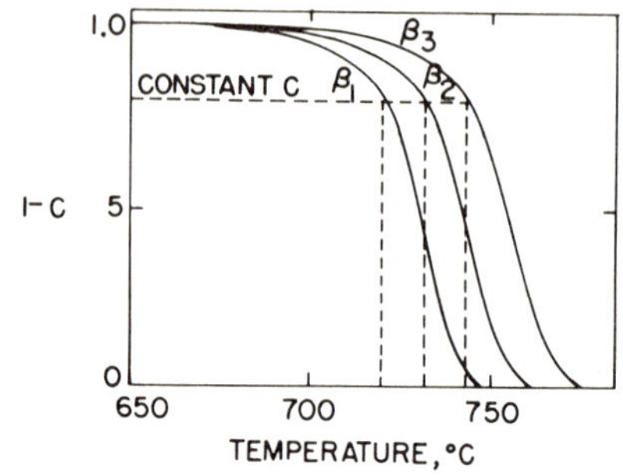

FIG. 5. Multiple thermogram kinetic plot.

activation energy may be calculated by Eq. (7). This procedure may be repeated at other values of 1 – C, thus testing the constancy of E with respect to C and T.

Friedman [12] suggested that if Eq. (8)

$$-\frac{d\gamma}{dT} = \frac{K}{b} \exp(-E/RT)\gamma^n \tag{8}$$

is arranged into the following form

$$\log [b(-d\gamma/dT)] = -\frac{E}{2.303R}\frac{1}{T} + n \log \gamma + \log K \tag{9}$$

a straight line should result when the log [b(–dγ/dT)] is plotted against 1/T for constant values of γ. The activation energy is determined from the slope, and the reaction order is determined from the intercept.

Another method is obtained from an approximate integral solution to Eq. (8). In this case,

$$\log\left(\frac{T^2}{b}\right) = \frac{E}{2.303R}\frac{1}{T} + \log [f(\gamma)E/RK] \tag{10}$$

A plot of log (T^2/b) vs. 1/T should yield a straight line for constant values of γ, and from the slope of this line the activation energy can be determined. However, it is not possible to obtain the reaction order or frequency factor using the above equation.

It should be expected that any analysis involving more than one data set, as is the case with multiheating rates, should improve the accuracy with which the kinetic parameters are determined. This is true provided that there is no change in the decomposition mechanism at the different heating rates. If there is a change in the mechanism, then erroneous conclusions can be drawn if Eq. (9) or Eq. (10) above is used in determining the kinetic parameters. Should multiheating rate data be available, the analysis of each thermogram should be incorporated into the study to determine the consistancy of the data.

A more sensitive method for obtaining the kinetic parameter of thermal decomposition employs the ratio method developed by Michelson and Einhorn [13].

A simple model equation is the form:

$$(-d\gamma/dt) = K \exp(-E/RT)\gamma^n \tag{11}$$

From Eq. (11) it is seen that $(-d\gamma/dt)$ and γ can be determined at any temperature, t, so that

$$(-d\gamma/dt)_i = K \exp(-E/RT_i)\gamma_i^n \tag{12}$$

At any other temperature, the values of $(-d\gamma/dt)$ and γ can also be determined. Using the subscript "j" to denote this case, the following equation applies:

$$(-d\gamma/dt)_j = K \exp(-E/RT_j)\gamma_j^n \tag{13}$$

By the logarithms of the ratios of Eqs. (12) and (13) the following useful equation can be derived:

$$\log [(-d\gamma/dt)_j/(-d\gamma/dt_i)] = \frac{E}{2.303R} \frac{T_j - T_i}{T_i T_j} - n \log \gamma_i/\gamma_j \tag{14}$$

The variables $\log (-d\gamma/dt)_j/(-d\gamma/dt)_i$ and $(T_j - T_i)/T_iT_j$ will plot as a straight line providing the ratio γ_i/γ_j is maintained constant. For example, if the ratio is set at 2, then

$$\gamma_i/\gamma_j = 0.90/0.45 = 0.80/0.40 = 0.70/0.35 \tag{15}$$

fulfills this condition and the slopes and temperatures are evaluated at each of these points. The activation energy can be determined from the slope of the line and the reaction order can be determined from the intercept.

The Isothermal Method

In reactions of the type (a), the onset of reaction involves the formation of the new phase B. Decomposition will begin and will spread out from the points (or nuclei) where the reaction was initiated. This initiation of nuclei formation in the reactant is called the induction period of the reaction and may be heat induced. The general form of the decomposition curve can be explained qualitatively in terms of the nuclei formation and growth. Other important phases of the reaction include the main portion of the reaction where the fractional weight loss is proportional to time, and the final decay period (which in the case of polymers may be masked by char formation). The rate equation applicable for reactions of this type is given by:

$$d\gamma/dt = K_{\gamma}{}^{n} \qquad (16)$$

where K is the probability of nuclei chain branching, and is assumed to be constant; γ is the fraction decomposed at time t; and n is the reaction order; $d\gamma$ is the incremental weight loss.

Then $$\Delta\gamma/\Delta t = K_{\gamma}{}^{n} \qquad (17)$$

and $$\log(\Delta\gamma/\Delta t) = \log K + n \log \gamma \qquad (18)$$

Table 8 illustrates the kinetic data obtained from a single isothermal thermogram (TGA).

A plot may be made of $-\log \gamma$ vs. $-\log(\Delta\gamma/\Delta t)$ (Fig. 6). Note: Negative values are used in order to obtain a positive slope. These plots may be made

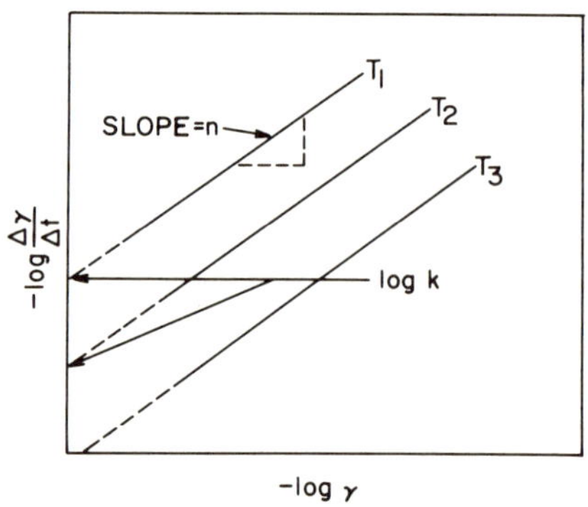

FIG. 6. Isothermal decomposition (TGA).

for isothermal determinations at several temperatures, as shown in Fig. 7 [14]. The slope is equal to n, the intercept is equal to log K. From the Arrhenius equation:

$$K = A \exp(-E/RT) \qquad (19)$$

Table 8. Data Obtained from a Single Isothermal Thermogram (Urethane Foam)

γ	$\log \gamma$	% wt loss	t	Δt	$\frac{\Delta\gamma}{\Delta t}$	$\log \frac{\Delta\gamma}{\Delta t}$
0.85	0.070	0.9	5.7			
				2.5	$\frac{0.1}{2.5} = 0.04$	1.398
0.75	0.125	0.8	8.2			
				2.8	$\frac{0.1}{2.8} = 0.036$	1.445
0.65	0.187	0.7	10.0			
				1.4	$\frac{0.1}{1.4} = 0.072$	1.145
0.55	0.260	0.6	11.4			
				1.1	$\frac{0.1}{1.1} = 0.091$	1.042
0.45	0.347	0.5	12.5			
				0.9	$\frac{0.1}{0.9} = 0.11$	0.955
0.35	0.455	0.4	13.4			
				1.0	$\frac{0.1}{1.0} = 0.10$	1.000
0.25	0.602	0.3	14.4			
				0.9	$\frac{0.1}{0.9} = 0.11$	0.955
0.15	0.825	0.2	15.3			
				1.3	$\frac{0.1}{1.3} = 0.07$	1.115
0.05	–	0.1	16.6			

where K is the rate constant, A is the frequency factor, and E is the activation energy, the other important parameters can be calculated.

Equation (18) may be written in logarithmic form:

$$\ln K = \ln A^{-(E/RT)} \qquad (20)$$

or
$$\log K = \log A - \frac{E}{2.303R} \tag{21}$$

Figure 8, the Arrhenius activation energy plot of the polymer decomposition, permits determination of log A (the y intercept) and the activation energy, E.

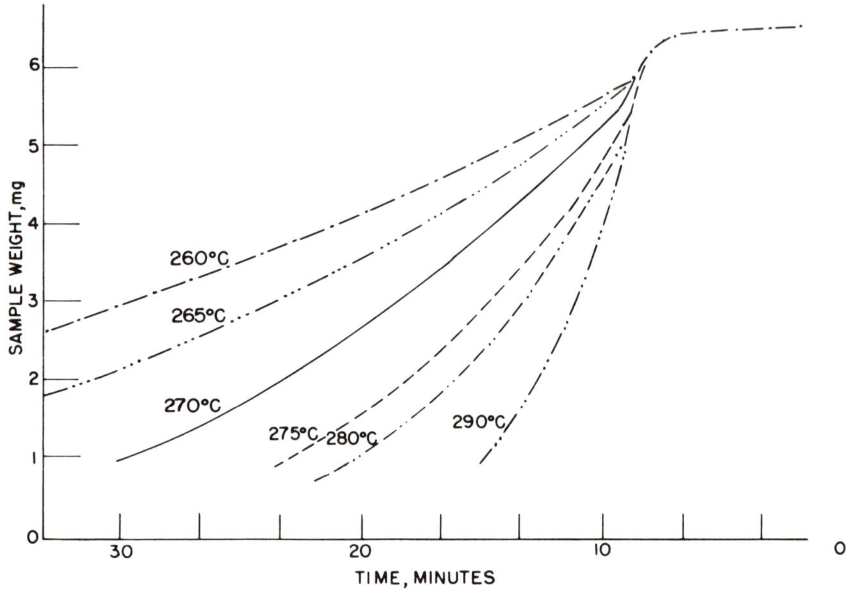

FIG. 7. Isothermal degradation of a urethane polymer.

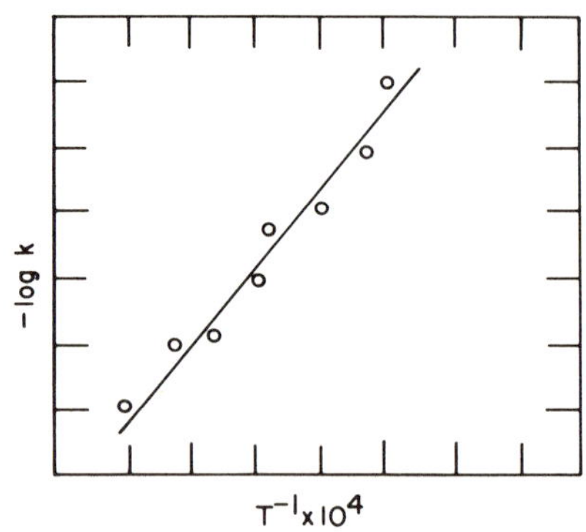

FIG. 8. Arrhenius activation energy plot of polymer decomposition.

The activation energy, E, may be determined from the relationship given in Eq. (22).

$$\text{Slope} = n = \frac{E}{2.303R} \tag{22}$$

For a constant value of γ, a plot of log ($-d\gamma/d\theta$) vs. 1/T should yield straight lines and the activation energy can be determined from the slope of the lines. Once the reaction order and the activation energy have been determined, the frequency factor can be determined from the logarithmic form of Eq. (11).

MEANS OF IMPARTING FLAME RESISTANCE TO PLASTICS

The design of flame resistance into plastics is essentially based on empirically gained knowledge. In practical form (that is, taking into account economic limitations and performance characteristics) flame resistance is imparted by the incorporation of halogen-bearing compounds, specifically chlorine and bromine, by phosphorus- or boron-containing compounds, and by nitrogenous materials, either singly or used in combination. Their effectiveness can be, and frequently is, augmented by auxiliary synergistically acting compounds.

Flame retardants can be classified into three categories:

1. Reactive type – The fire retardant reacts chemically with the polymer components and is combined chemically into the basic polymeric structure.
2. Additive type – The fire retardant is physically blended into the base polymer prior to conversion to the final plastic product.
3. Coadditive type – Includes compounds which by themselves are ineffective as flame retardants, but in the presence of halogen-, phosphorus-, nitrogen-, or boron-containing compounds provides synergistically improved flammability characteristics.

As a general rule, it is found to be desirable to incorporate those compounds containing the flame-resistant agents as an integral part of the polymer molecule. By doing so, the plasticizing effects of relatively low molecular weight compounds are minimized; hence, thermal stability, dimensional stability, physical strength, and, in particular, the permanence aspect of flame resistance is enhanced.

The actual selection of fire-retardant type is largely dependent on the type of polymer, method of preparation, and end-use application. Reactive-type fire retardants are normally used with thermosetting polymeric systems, while the additive-type fire retardants are usually preferred for thermoplastic polymers.

Mechanism of Halogenated Fire Retardents

Halogenated fire retardants are thought to act through one of two main mechanisms: In one, the fire retardant causes the burning substrate to follow an abnormal course of decomposition, yielding less flammable gaseous products. In the other mechanism, the fire retardant itself decomposes to gaseous products which mix with the normal gaseous products from the decomposing polymer to yield a less flammable gas mixture. Skinner [15], in a recent review of the literature, summarized the role of halogenated fire retardants in quenching burning polymers. The normal oxidation chain reaction is shown in Eq. (23).

$$\begin{array}{lcl} {}^{\bullet}OH + RCH_3 & \longrightarrow & RCH_2{}^{\bullet} + H_2O \\ RCH_2{}^{\bullet} + O_2 & \longrightarrow & RCHO + {}^{\bullet}OH \end{array} \qquad (23)$$

The ˙OH radical attaches a hydrocarbon (polymeric material) and generates a hydrocarbon radical and water. In the presence of oxygen the hydrocarbon regenerates the ˙OH radical. Some of the gases generated during the decomposition may be flammable themselves, but, when mixed with other flammable gases from the decomposing plastics, they produce a gas-air mixture that is too rich to burn. The chemical explanation is shown in Eq. (24).

$$R'H + {}^{\bullet}OH \longrightarrow \underset{\text{(less reactive than }{}^{\bullet}\text{OH)}}{R'^{\bullet}} + H_2O \qquad (24)$$

The excess hydrocarbon acts as an oxidation suppressor, and thus the very reactive hydroxyl radicals which normally serve to maintain the oxidation chain reaction are used up by the hydrocarbon.

Halogen-containing fire retardants decompose to gases that form free radicals in the flame and these radicals can become chain terminators in the oxidative chain reactions of the flame. This is illustrated in Eq. (25) with bromine from hydrogen bromide.

$$\begin{array}{lcl} \text{Brominated flame retardant} & \xrightarrow{\Delta} & Br^{\bullet} \\ Br^{\bullet} + R'H & \longrightarrow & R'^{\bullet} + HBr \\ {}^{\bullet}OH + HBr & \longrightarrow & H_2O + Br^{\bullet} \end{array} \qquad (25)$$

When the concentration of the reactive free-radical intermediates is reduced, the flame reactions slow down and the flame extinguishes. Lash and Wagner [16] reported that the following compounds were found to decrease the flame speed of a hexane-air flame: Br_2, PCl_3, $POCl_3$, $(CH_3)_3PO_4$, $SbCl_3$, $SaCl_4$, $TiCl_4$, and $(CH_3\,CH_2)_4\,Pb$.

Piechota [17] reported that hydrogen chloride was formed via an intramolecular elimination from different chlorine-containing compounds. (Table 9).

Tilley [18] reported that sodium chloride and sodium bromide did not contribute to the protection of urethane foams during fire exposure. Presumably, neither halogen nor hydrohalide was available from salts during combustion. Both Piechota [17] and Friedman and Levy [19] have collected data which strongly suggest that for brominated retardants the formation or liberation of HBr, and especially bromine, during the combustion is necessary for retardance. With chlorine-containing retardants, the active moiety appears to be HCl. Aromatically bound halogen compounds seemed to perform more poorly than aliphatic halogen compounds, possibly because the arylhalogen bond is more difficult to cleave than an aliphatic halogen bond.

FLAME RETARDANCE CONFERRED BY A HALOGEN IN A REACTIVE INTERMEDIATE

Numerous investigators have reported on studies in which the flame-proofing elements such as a halogen or phosphorus or both are incorporated into a reactive intermediate. Li et al. [20] used pentaerythritol dichloride as the polyhydric alcohol moiety in preparing unsaturated polyester resins. A chlorine content of at least 25% was considered essential to obtain a desired level of flame retardancy.

Al'shits and co-workers [21] used a mixture of pentaerythritol dichloride and trichloride to flame retard unsaturated polyester resins. The trichloride was used to form end groups in the linear chains thus raising the molecular weight of the polymer (as compared to previous studies using only the dichloride) while also raising the halogen level in the polyester. These polyester resins were prepared using a two-stage condensation reaction. In the first stage excess maleic and phthalic anhydride were reacted with pentaerythritol dichloride. Sufficient pentaerythritol trichloride was then added so that the total hydroxyl groups in the system would predominate and the condensation reaction would be completed.

Resorcinol has been used as the starting material to prepare halogen-

Table 9. Intramolecular Elimination of HCl from Different Chlorine Compounds

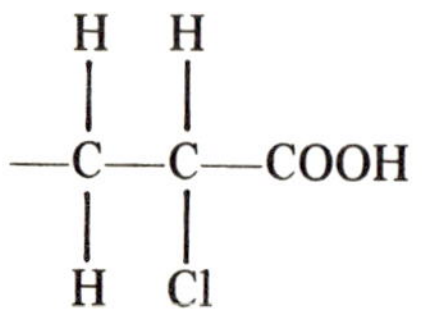	→	—CH = CH – COOH + HCl
Chlorinated fatty acid		unsaturated acid
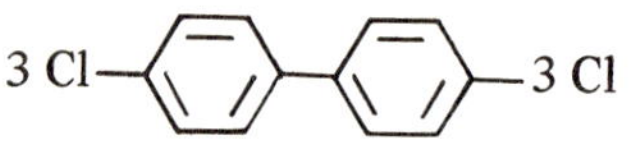	→ −HCl	nonaromatic compound
Chlorinated diphenyl		
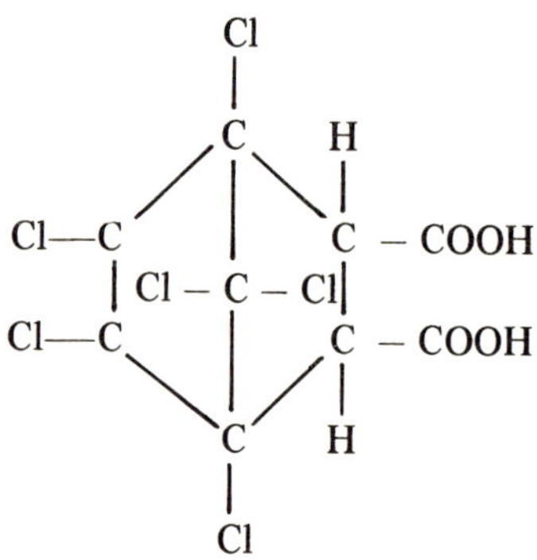	→	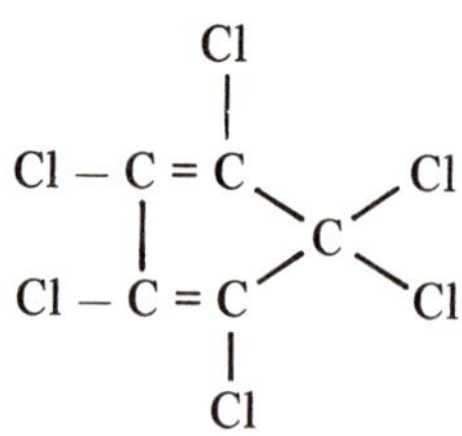
HET acid		Hexachlorocyclopentadiene + 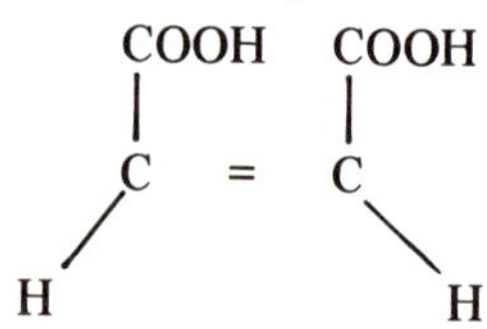 maleic acid
NH_4CL ammonium chloride	→	NH_3 + HCl

containing diols (22,23). The resorcinol was epoxylated with ethylene or propylene oxide to yield 1,3-bis(β-hydroxyalkoxy) benzenes. These were then brominated in a buffered solution to form B,B′-(4,6-dibromo-1,3-phenylenedioxy) dialkanols which were used to prepare self-extinguishing polyesters [Eq. (26)].

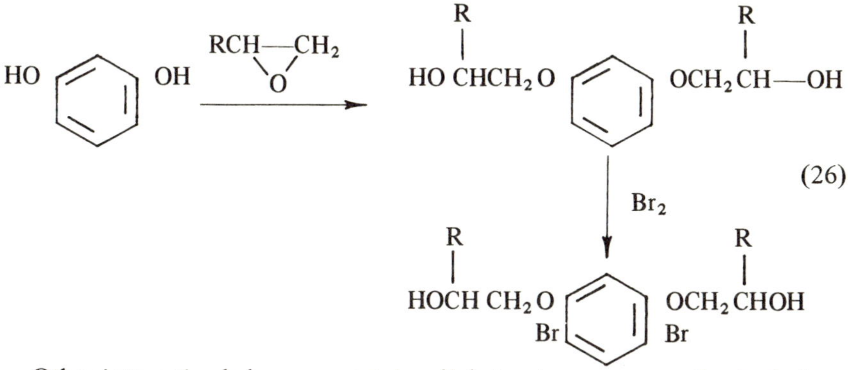

(26)

Other interesting halogen-containing diol structures were synthesized via a Diels-Alder reaction of hexahalocylopentadienes with unsaturated diols (such as 1,4-butenediol) used as the dienophiles [24,25]. Replacement of some of the chlorine atoms by bromine in the diene increases the flame-retardance properties in polyesters, while replacement of some of the chlorine atoms with fluorine yields polyesters with enhanced aging and heat-resistant properties.

Recently, considerable interest has been shown in polyester resins based on tetrachlorophthalic anhydride and tetrabromophthalic anhydride [26-30].

Laboratory studies were conducted to evaluate the chemical and physical parameters which govern the flammability characteristics of these retardants in model unsaturated polyester resin systems.

The highest saturated-to-unsaturated molar ratio consistent with good resin properties was selected in order to permit synthesis of polyesters with the highest possible chlorine content. A ratio of saturated-to-unsaturated acid of 1.1 to 1.0 was chosen since higher ratios of tetrachlorophthalic anhydride to maleic anhydride would cause a rapid drop in heat distortion temperature. This ratio was held constant in all model systems by adjusting the amount of phthalic anhydride used. Propylene glycol was selected as the glycol for this study and a 5% molar excess was used in all formulations. The resins were prepared by a conventional fusion esterification with the reaction temperature being maintained at 190°-195°C. The resins were processed to an acid number of 30, styrenated to a level of 25 or 30%, and inhibited with 125 ppm of hydroquinone. Table 10 presents the formulations used to prepare resins based on tetrabromophthalic anhydride. Table 11 shows the formulations used to

Table 10. Formulations for Polyester Resins Based on Tetrabromophthalic Anhydride

Moles TBPA	Moles PA	Moles MA	Calculated % bromine on dilution with 25% Styrene	Calculated % bromine on dilution with 30% Styrene
0.80	0.30	1.0	30.4	–
0.50	0.60	1.0	22.0	20.5
0.30	0.80	1.0	14.8	13.8
0.25	0.85	1.0	12.7	11.8
0.23	0.87	1.0	11.8	11.0
0.20	0.90	1.0	10.5	9.8
0.07	1.03	1.0	5.0	–

prepare resins based on tetrachlorophthalic anhydride. Figures 9 and 10 present the idealized structures of cross-linked polyester based polymers containing the halogenated fire retardants.

Table 11. Formulations for Polyester Resins Based on Tetrabromophthalic Anhydride

Moles TCPA	Moles PA	Moles MA	Calculated % chlorine on dilution with 25% Styrene	Calculated % chlorine on dilution with 30% Styrene
1.0	1.0	1.0	21.7	20.3
1.0	0.1	1.0	20.2	18.9
0.9	0.2	1.0	18.7	17.4
0.8	0.3	1.0	17.0	15.8
0.5	0.6	1.0	11.5	–
0.2	0.9	1.0	5.0	–

The effect of fire retardants on the thermal characteristics of model polyester resin systems was determined using differential thermal analysis and thermogravimetric analysis. Figures 11 and 12 illustrate the effect of fire-retardant concentration on the decomposition temperature for specimens fire retarded with the two anhydrides.

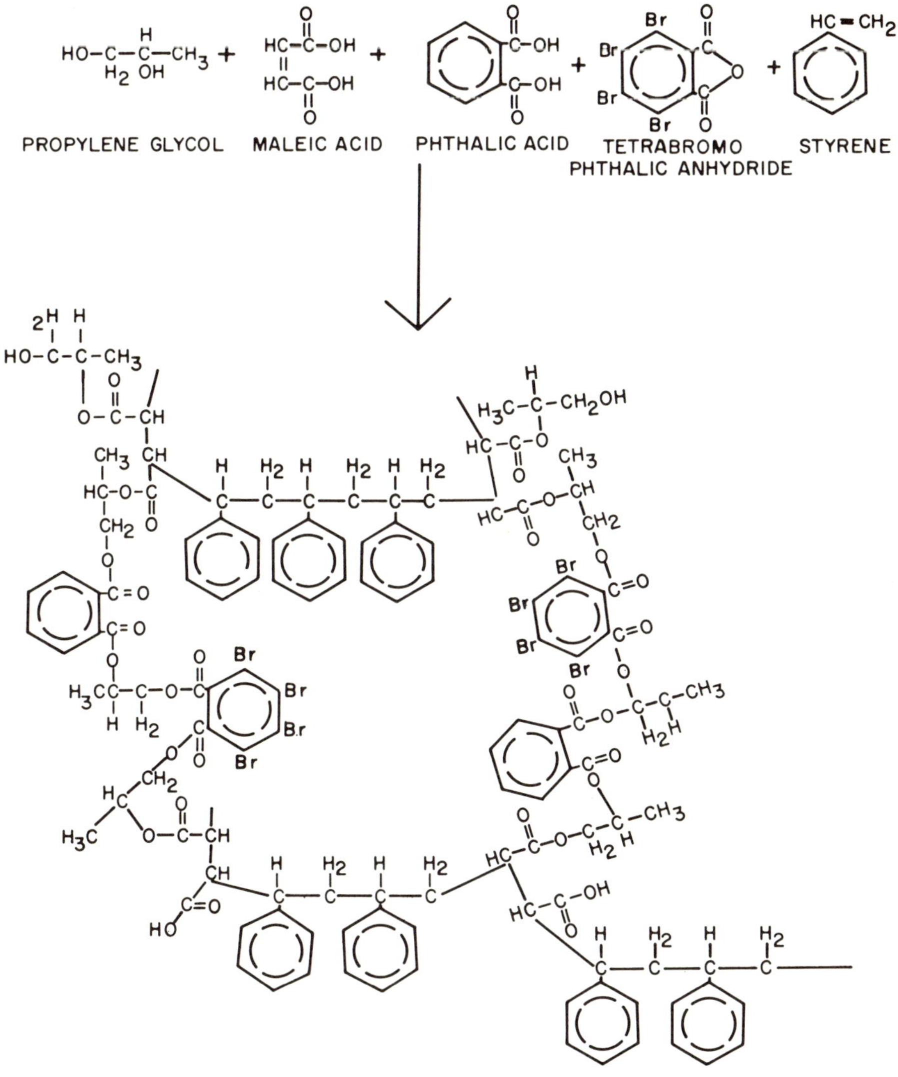

FIG. 9. Idealized structure of a cross-linked polyester-based polymer containing tetrabromophthalic anhydride.

Relatively little difference was observed in the decomposition temperatures for the series containing 25% styrene and the series containing 30% styrene.

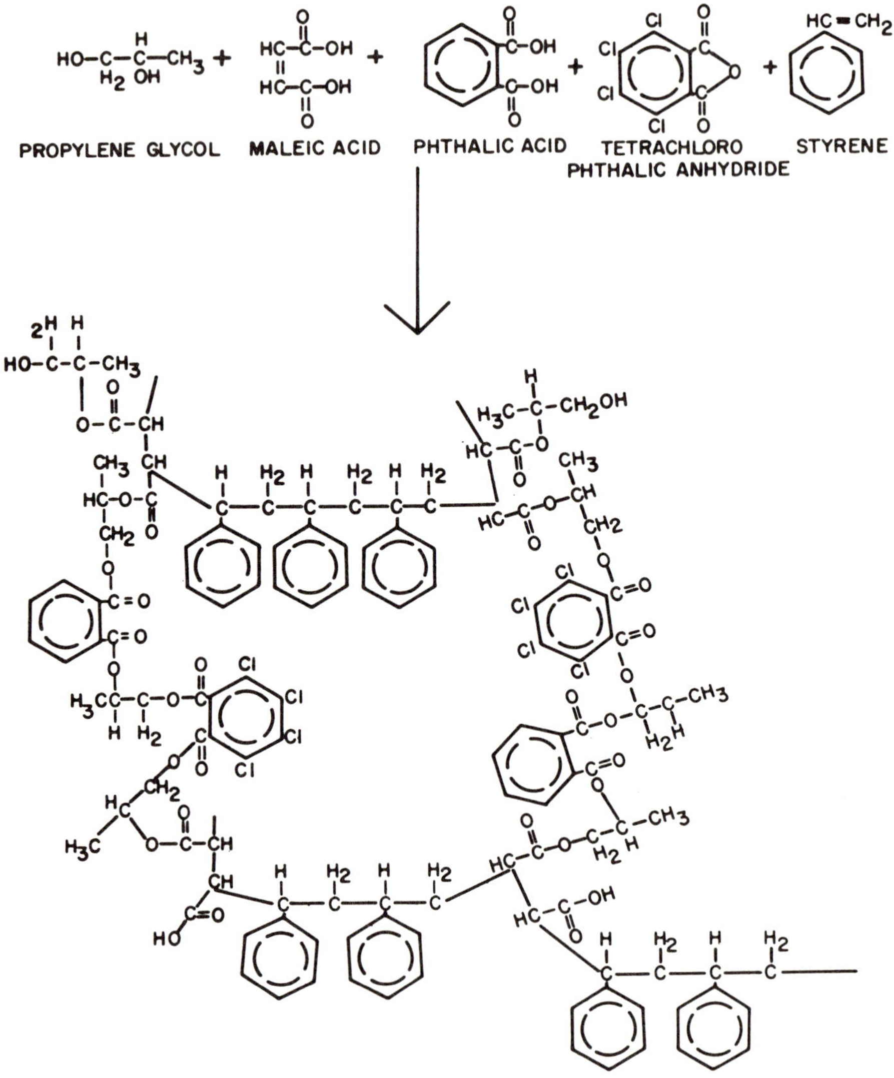

FIG. 10. Idealized structure of a cross-linked polyester-based polymer containing tetrachlorophthalic anhydride.

The control samples containing only the halogenated fire retardant showed a slight lowering of the decomposition temperature. The incorporation of

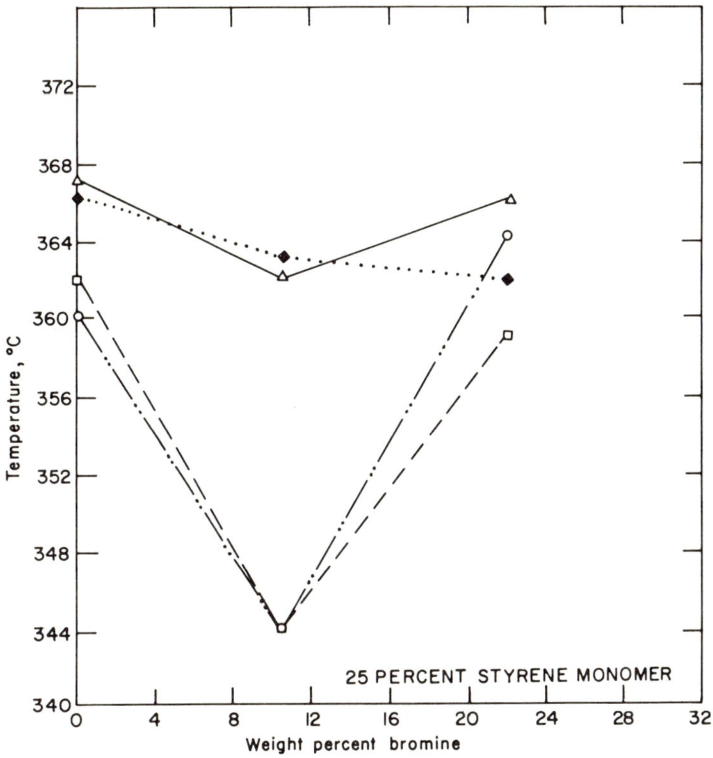

FIG. 11. The effect of tetrabromophthalic anhydride on polymer decomposition: (· · ·) no additive, (—) 3% TEP, (– · · ––) 2% Sb_2O_3, and (– – –) 5% Sb_2O_3.

triethyl phosphate into these halogenated systems showed no marked synergistic effects. Incorporation of antimony trioxide (2% or 5% by weight) resulted in a substantial lowering of the decomposition temperature with an observed minimum occurring at approximately 11% bromine concentration or at approximately 21.7% chlorine content.

Two small-scale test methods were employed under controlled laboratory conditions to evaluate the relative effectiveness of the polyesters modified by the halogenated fire retardants. Figures 13 and 14 show the apparatus used for the ASTM D-757 and the HLT-15 [31] test methods.

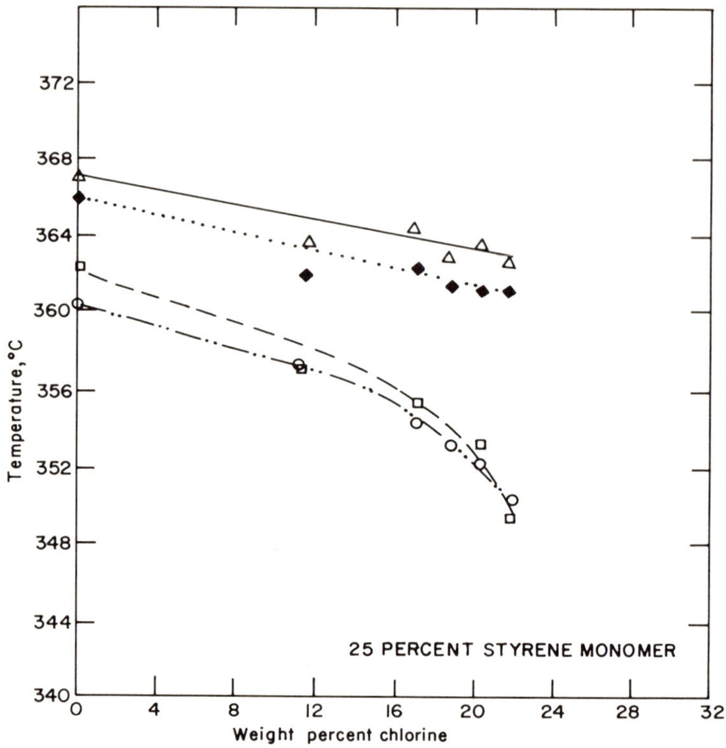

FIG. 12. The effect of tetrachlorophthalic anhydride on polymer decomposition. Key same as Fig. 11.

ASTM D-757 Globar Test

In this test a Globar element is heated to 950°C and a specimen 5 by 1/2 by 1/8 in. is brought into contact with the bar for 3 min. After this period, the flame, if any, is extinguished. The distance burned is measured and the rate of flame propagation is reported in inches per minute. Highly flame-resistant specimens usually char and burn for only a fraction of the ignition period. Figures 15, 16, and 17 illustrate the results obtained when the fire-retardant test specimens were evaluated by the ASTM D-757 test method. It is obvious that at equal concentrations of halogen, the specimens based on tetrabromophthalic anhydride have lower flame propagation values than were observed for similar specimens based on tetrachlorophthalic anhydride. The incorporation of

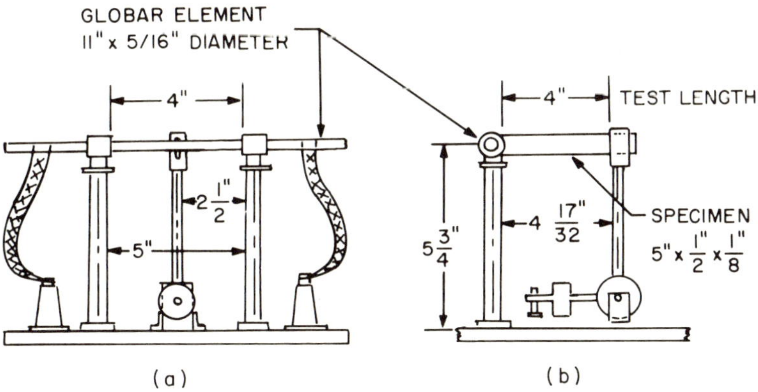

FIG. 13. ASTM D-757 apparatus: (a) front view and (b) side view.

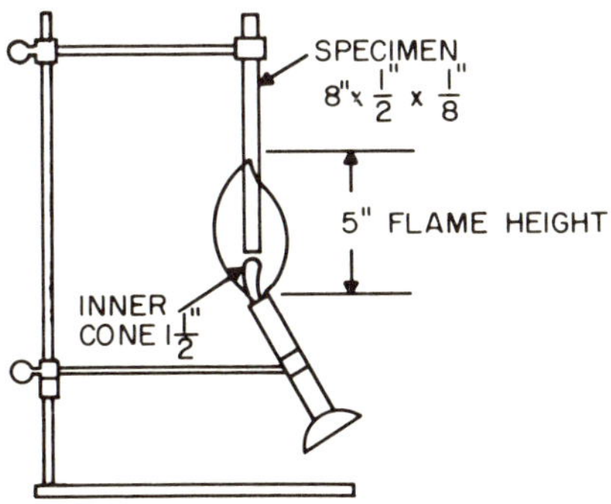

FIG. 14. HLT-15 apparatus.

triethyl phosphate reduces the burning rate slightly. The incorporation of antimony trioxide exhibits a strong synergistic effect and markedly improves the flame propagation characteristics of the test specimens.

HLT-15 Intermittent Flame Test

This test is more severe than the ASTM D-757 test previously described because the specimen is suspended in a vertical position and heat from the burning will be carried upward by convection along the length of the specimen. In addition, the specimen is ignited five times using increasingly longer ignition periods. The sequence of ignition times and flame withdrawals are given in the accompanying tabulation.

Flame application	Ignition time, sec	Burner withdrawn, sec
1	5	10
2	7	14
3	10	20
4	15	30
5	25	50

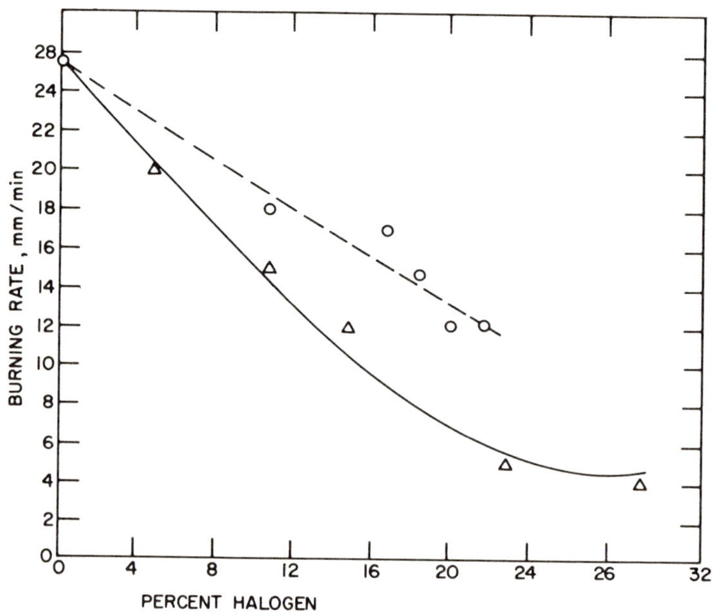

FIG. 15. ASTM D-757 flame test, no additive: (---) chlorine and (—) bromine.

In rating a composition, fire specimens 8 by 1/2 by 1/8 in. are burned. If the flame is extinguished within the period that the burner flame is withdrawn, the specimen has passed the ignition test. Each of the five test ignitions successfully passed is worth 20 points toward the rating for that specimen. An average is taken of the ratings for the five specimens. Figures 18, 19, and 20

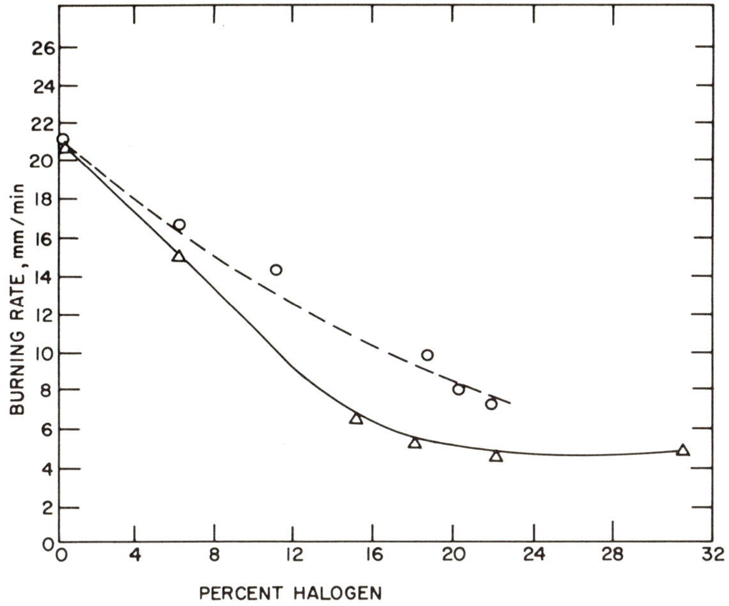

FIG. 16. ASTM D-757 flame test, 3% TEP: (– – –) chlorine and (—) bromine.

show the results obtained when the test specimens based on the halogenated anhydrides were evaluated by the HLT-15 test method. The brominated specimens obtained a maximum rating without the addition of additional co-retardants. Slight improvements were seen with the addition of triethyl phosphate. The chlorinated specimens were not able to exhibit burning characteristics to obtain greater than a 42 rating by this test method. Incorporation of antimony trioxide improved the ratings of both the brominated and chlorinated specimens, but the brominated specimens were more effectively fire retarded than the chlorinated specimens.

FIRE RETARDANCE CONFERRED BY A HALOGEN IN A NONREACTIVE INTERMEDIATE

Nonreactive fire retardants are generally used with thermoplastic polymeric systems where incorporation of a reactive material would affect crystallinity, heat distortion temperature, and resistance to ultraviolet light or humidity. Low molecular weight fire retardants have a tendency to migrate or outgas during processing or while in actual use. Thus, consideration should be given

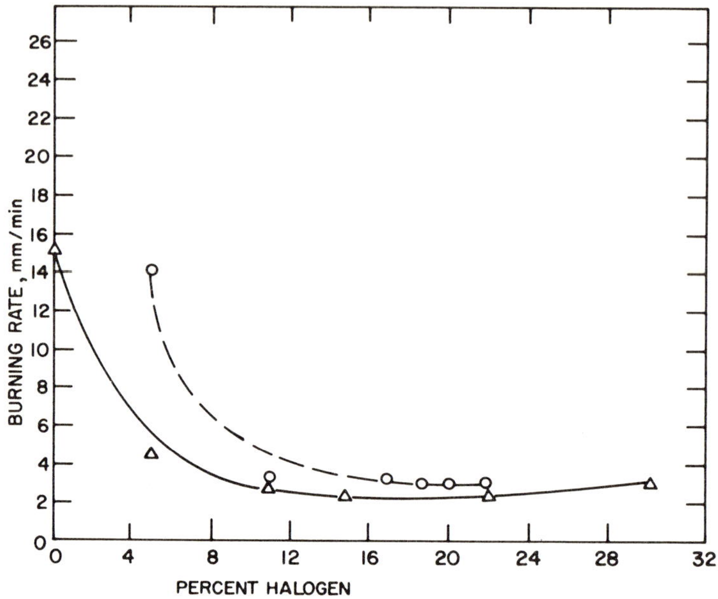

FIG. 17. ASTM D-757 flame test, 5% Sb_2O_3: (– – –) chlorine and (—) bromine.

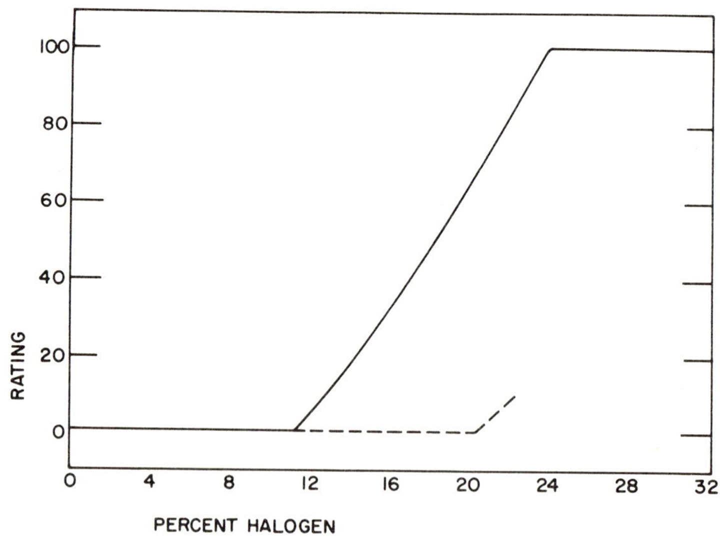

FIG. 18. HLT-15 flame test, no additive: (– – –) chlorine and (—) bromine.

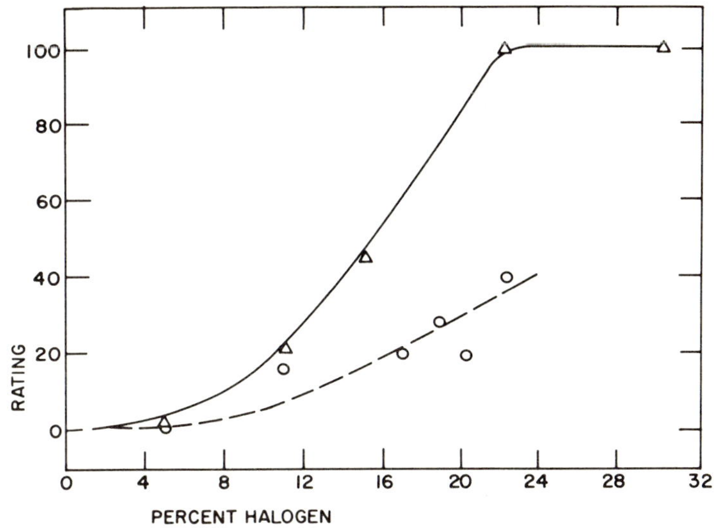

FIG. 19. HLT-15 flame test, 3% TEP: (– – –) chlorine and (—) bromine.

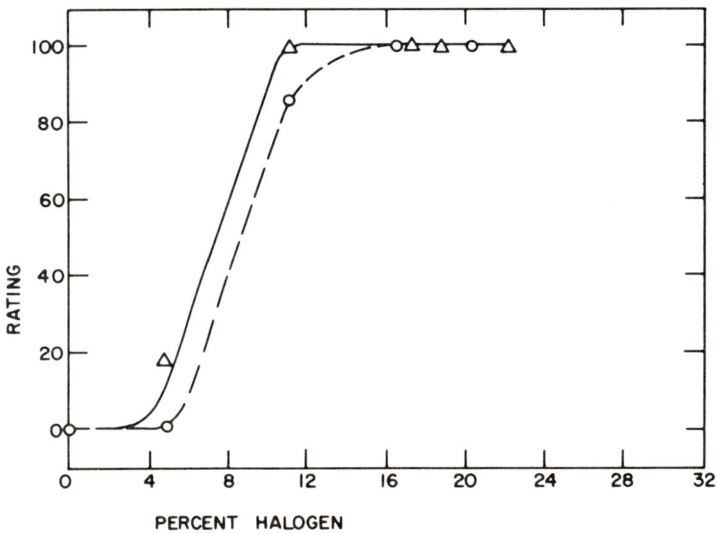

FIG. 20. HLT-15 flame test, 5% Sb_2O_3: (– – –) chlorine and (—) bromine.

to those compounds having a molecular weight high enough to prevent sublimation, migration, or volatilization.

Selection of halogenated nonreactive fire retardants must be based on chemical structure, especially in regard to the nature of the attached hydrocarbon moiety. Typical structural effects that may influence the effectiveness of the flame retardancy characteristics of halogenated hydrocarbons are in saturated compounds the differences between primary, secondary, and tertiary locations. In unsaturated compounds the bond strength of halogens in allyl, benzyl, vinyl, and aryl structures will vary widely. The location of nearby polar or nonpolar groups may affect the behavior of the carbon-halogen bonds. Alkylhalogens are easily lost when subjected to high temperatures while arylhalogens exhibit greater thermal resistance.

PHOSPHORUS-HALOGEN SYNERGISM

Synergism is the term applied to the observed ability of two materials to influence the properties or response of a material to a greater degree than equal amounts of the two materials used separately.

Mitchell and Murphy [32] showed the results obtained when urethane foams were fire retarded with phosphorus- and halogen-containing fire retardants (Table 12).

Table 12. Effect of Phosphorus and Chloride Inhibitors on the Autoignition Temperature and Char Strength of a Foam

Inhibitor Phosphorus content, %	Chloride content, %	Autoignition, °F	Char strength
0	0	950	None
0	50	950	None
5	50	1000	Weak
10	0	1150	Strong

The foams retarded with halogen-containing compounds were totally consumed without the development of char structure, while those retarded with phosphorus- and halogen-containing compounds showed both a quenching response to fire exposure and the development of a carbonaceous char structure.

Hilado (33) explained the mechanisms employed by phosphorus and bromine compounds as follows: ". . . without phosphorus, hydrogen bromide is believed to be the most active bromine compound, since this compound boils in the combustion zone, where it participates in successive halogenation and dehydrohalogenation reactions. The presence of phosphorus promotes the formation of char, which further restricts movement in the gaseous phase, and results in the formation of phosphorus tribromide, phosphorus pentabromide, and phosphorus oxybromide, which are less readily gasified and are heavier gases (molecular weights of 270.70, 430.52, and 286.70, respectively). Since these bromides are continually regenerated in the halogenation-dehydrohalogenation process, their effectiveness increases with their residence time in the combustion region, and keeping them in effective control twice as long would tend to make them twice as effective."

The fluxing action of phosphorus is probably limited to the interface of the combustion and pyrolysis regions so that any excess over the amount needed to stabilize the char is ineffective. Since phosphorus oxides can effect only so much dehydration while the continually regenerated bromides can repeat the dehydrohalogenation process, phosphorus-bromine combinations would be more effective than phosphorus oxides in promoting char formation.

The mechanism of the phosphorus-chlorine synergism is similar to that of the phosphorus-bromine synergism. It is more widely utilized because of the greater availability of chlorine-containing compounds. The phosphorus-chlorine synergism is less effective than the phosphorus-bromine synergism to the same degree and for the same reasons that chlorine is not as effective on an equal weight basis as bromine. The chlorides of phosphorus are lower boiling and lighter gases than the corresponding phosphorus bromides and can be expected to have a shorter residence time in the combustion zone.

ANTIMONY-HALOGEN SYNERGISM

Various antimony compounds have been used as flame retardants for natural and synthetic polymers [34-37]. Antimony trioxide (Sb_2O_3) and antimony oxychloride (SbOCl) are the most commonly used antimony compounds for modifying the flammability characteristics of polymeric materials. Schmidt [38] and Drake [39] reported that antimony trioxide by itself was a very poor flame retardant. Little [35] observed that the optimum flame retardancy occurred when the mole ratio of antimony to chlorine is 1:1.

The HCl evolved from the decomposition of chlorinated organic compounds reacts with antimony oxide to produce the trichloride (bp 223°C) or the oxychloride, which in turn decomposes to the trichloride. The strongly acidic trichloride is thought to function as a dehydration catalyst for materials such as cellulose as well as a flame quencher for flammable gases. Figure 21 summarizes the thermal reactions of antimony compounds.

$$Sb_2O_3 + 6HCl \longrightarrow 2\,SbCl_3 + 3H_2O \quad \text{(bp 223 °C)}$$

$$Sb_2O_3 + 2HCl \longrightarrow 2\,SbOCl + H_2O$$

$$5\,SbOCl \xrightarrow{170-280°C} SbCl_3 + Sb_4O_5Cl_2$$

$$3\,Sb_4O_5Cl_2 \xrightarrow{320-700°C} 2\,SbCl_3 + 5Sb_2O_3$$

FIG. 21. Thermal reactions of antimony compounds.

Pitts et al. [40] found the thermal decomposition of SbOCl to proceed in four basic steps as illustrated in Fig. 22.

$$5\,SbOCl_{(S)} \xrightarrow{245°-280°C} Sb_4O_5\,Cl_{2(S)} + SbCl_{3(g)}\uparrow$$

$$4\,Sb_4O_5Cl_{2(S)} \xrightarrow{410°C-475°C} 5\,Sb_3O_4Cl_{(S)} + SbCl_{3(g)}\uparrow$$

$$3\,Sb_3O_4Cl \xrightarrow{475°C-565°C} 4\,Sb_2O_3 + SbCl_{3(g)}\uparrow$$

$$Sb_2O_{3(S)} \xrightarrow{658°C} Sb_2O_{3(l)}$$

FIG. 22. Thermal decomposition of antimony oxychloride.

Einhorn et al. [41] have shown the effectiveness of the antimony-halogen synergistic reaction in reducing the tendency for crack growth and enhancing the development of a homogeneous char structure in unsaturated polyester resins through the use of infrared photographic techniques. Figure 23 shows

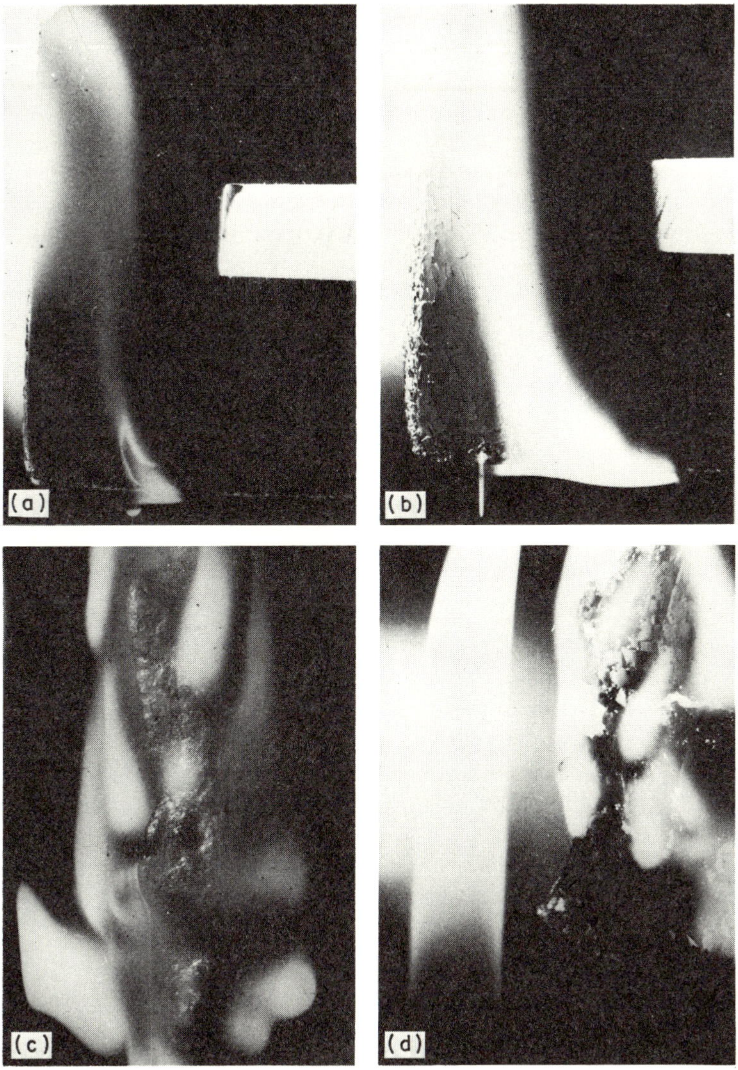

FIG. 23. Degradation of a polyester resin casting fire retarded by incorporation of 11.5% by weight of tetrabromophthalic anhydride: (a) 1 sec, (b) 5 sec, (c) 60 sec, and (d) 90 sec.

the degradation of a polyester resin casting fire retarded by the incorporation of 11.5% by weight of tetrabromophthalic anhydride. Figure 23a illustrates the melting region, the region of char formation, and also shows flame propagation by burning gases in the boundary region. Figure 23b shows the development of a crack which has started in the boundary region, char sintering, and afterglow (cooling). Figure 23c shows the extension of the melting zone and further crack growth. Figure 23d shows char sintering and flame propagation behind the weak nonhomogeneous char front.

A polyester resin casting fire retarded with 5% antimony oxide and 11.5% tetrabromophthalic anhydride was subjected to burning (Fig. 24). Figure 24a shows the development of a strong char structure in the flame front. In contrast to the plastic fire retarded with only the halogen-containing retardant, there is no evidence of flame propagation beyond that area in direct contact with the flame source by burning gases in the boundary layer. Figure 24b shows the gradual erosion of the strong homogeneous char structure in the area of actual flame contact. Figure 24c shows rapid extinguishment of the flame front after travel beyond the initial flame source. Figure 24d illustrates the char structure after flameout.

BORON FIRE RETARDANTS

The largest use of boron compounds as fire retardants is in wood and wood products. Eickner [42] published a comprehensive review covering the performance characteristics of fire-retardant wood. Borax, boric acid, and sodium borate are the most common boron compounds used in treatment of wood.

McCluer [43] tested a number of inorganic boron salts as fire retardants for fabrics. The most effective of those compounds evaluated was sodium fluoborate.

In general, organoboron derivatives (esters) are thought to be too hydrolytically unstable for general use as fire retardants. Knaggs [44] found that solutions of triamyl borate in an organic solvent with tritolyl phosphate and a fatty acid salt were effective in fire-retarding fibrous materials.

Cyclic boric acid esters derived from diethanolamines are stabilized toward hydrolysis by an intramolecular, boron-nitrogen coordinate bond

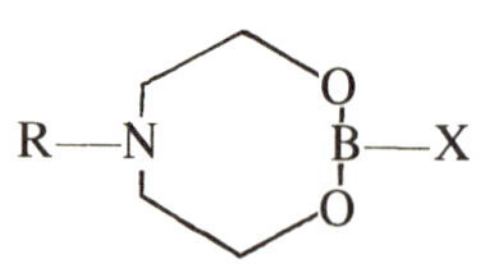

Rudner and Moores [45] utilized this structure to prepare direct polymers with cellulose that do not leach out on exposure to water.

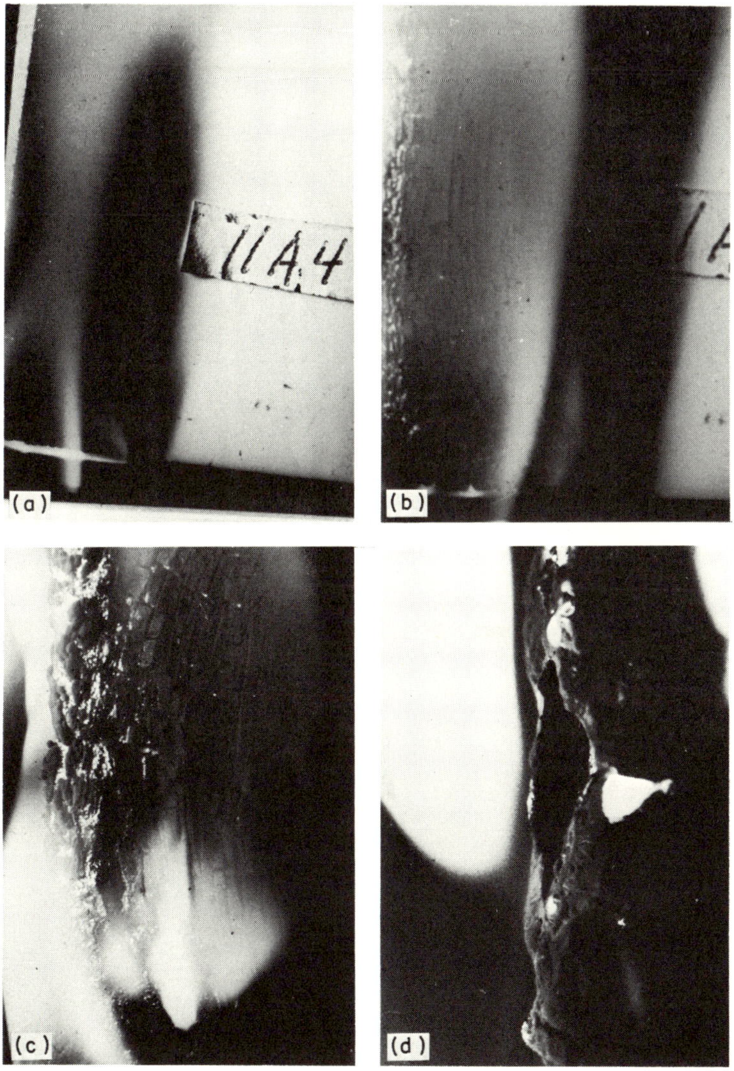

FIG. 24. Degradation of a polyester resin casting fire retarded with 5% antimony oxide and 11.5% tetrabromophthalic anhydride: (a) 1 sec, (b) 5 sec, (c) 60 sec, and (d) 90 sec.

Cellulose fibers when treated with the following boron-containing compound were appreciably slower burning than control fabrics.

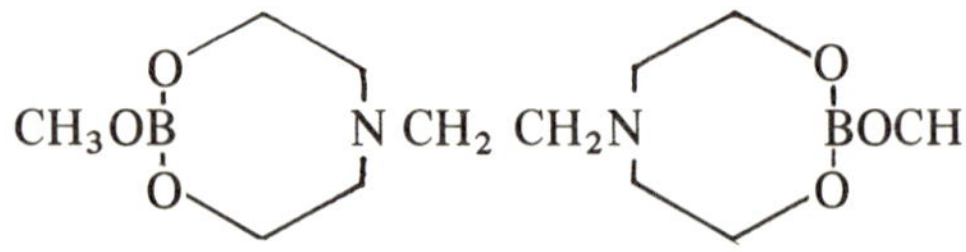

Rudner and Moores assumed that the methoxy groups were displaced by the hydroxyl groups of the cellulose to give an integrally bound boron-containing cross-linkage.

The mechanistic details of how boric acid and borate salts retard the combustion of cellulose are only partially understood. Schwenker and Beck [46] reported at least 37 products formed during the pyrolysis of cellulose. The major volatile primary product is thought to be levoglucosan (1,6-anhydro-β-D-glycopyranose). Levoglucosan is thought to arise by the thermal scission of the 1,4-glucoside bonds of cellulose.

A free primary hydroxyl group at C_6 is necessary for this depolymerization to occur. Levoglucosan is presumed to fragment to form the observed low molecular weight, volatile products. A second pathway for cellulose decomposition is simple dehydration to give water and dehydrocellulose [47]. The dehydrocellulose is believed to be the principal source of char.

Woods et al. [48] incorporated boron in the polyol moiety of rigid urethane foams. Several boron-containing compounds were synthesized as shown in Eqs. (27)-(30).

$$B(OH)_3 + 3CH_3CH(OH)CH_2OH \longrightarrow \underset{\text{(I)}}{B(OCH_2\underset{\displaystyle OH}{\underset{|}{C}}HCH_3)_3} + 3H_2O \qquad (27)$$

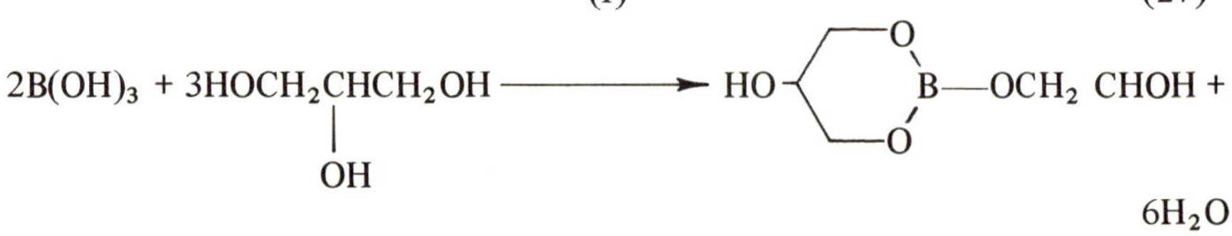

(II) (28)

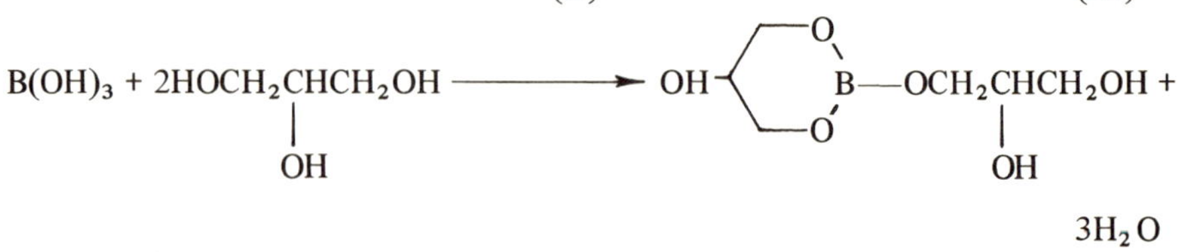

(III) (29)

$$B(OH)_3 + 3HO CH_2\underset{\displaystyle OH}{\underset{|}{C}H}CH_2OH \longrightarrow B(OCH_2\underset{\displaystyle OH}{\underset{|}{C}H}CH_2OH) + 3H_2O \qquad (30)$$

(IV)

Structures (I)-(IV) are idealized; the actual products are probably mixtures of boron-polyol polymers. Prepolymers prepared with structures (I), (IV), or tris-(hydroxypropyl) glycerine and toluene diisocyanate produced foams of such poor quality that their evaluation was discontinued. Foams were prepared using polymethylene polyphenylpolyisocyanate and the boron-containing polyols. Table 13 gives the formulations used by Woods et al. in preparing these foams.

The formation of a complex (IV) between the triethylenediamine catalysts and the acidic glycol borates resulted in retarded rates of rise during foam preparation.

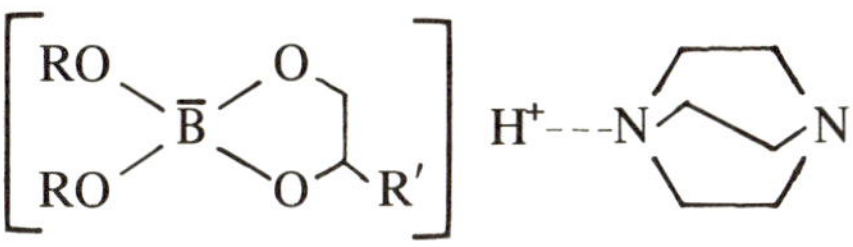

Additional amine catalyst was required to produce satisfactory foam specimens.

Table 14 presents the results of evaluations performed to determine the flammability characteristics of the boron-modified foams.

Woods et al. [49] prepared several brominated boric acid esters as potential nonreactive fire retardants. Rigid foams were prepared from a polyhydroxylpropylated sucrose-TDI prepolymer and polyether polyols. The additives were introduced at a level of 20 phr and the effect on burning rate determined. Some reductions in burning rates were achieved, but no self-extinguishing foams were obtained.

METHODS USED FOR THE FORMULATION OF FIRE-RETARDANT PLASTICS

The use of recently developed techniques of thermal analysis to match the chemical and physical properties of candidate fire retardants with that of the polymer to be stabilized has been reported by Einhorn [50]. Simultaneous differential thermal analysis (DTA), thermogravimetric analysis (TGA), and derivative thermogravimetric analysis (DTG) are carried out under dynamic conditions of a heating rate of 10°C/min. Figure 25 presents the thermograms for

Table 13. Polymethylene Polyphenylpolyisocyanate Boron-Polyol Foam Formulations[a]

Foam No.	PAPI, wt g	Polyol used	Polyol, wt g	Catalyst system	Catalyst, wt g	F-11, wt g
1	16.0	(I)	8.4	Triethylenediamine-hexanetriol[b]	0.4	6
2	16.0	(II)	19.1	Triethylenediamine-hexanetriol;	0.4	6
				$(C_2H_5)_3H$	0.2	
3	16.0	(III)	12.6	Triethylenediamine-hexanetriol;	0.4	6
				$(C_2H_5)_3N$	0.2	
4	13.1	(IV)	9.0	Triethylenediamine-hexanetriol	0.4	6
5	16.0	(IV)	9.3	Triethylenediamine-hexanetriol;	0.4	6
				$(C_2H_5)_3N$	0.2	
6	15.5	(IV)	9.3	Triethylenediamine-hexanetriol;	0.4	6
				$(C_2H_5)_3N$	0.2	
7	21.5	[e]	15.1	H,N,H′,N′-Tetramethyl-1,3-butane diamine	0.2	5

[a]Each foam formulation contained 0.2 g of DC-113 silicone surfactant (Dow Corning Co.).
[b]A 112 mixture of triethylenediamine (DABCO, Air Products & Chemicals, Inc.) and hexanetriol.
[c]Polymethylene polyphenylpolyisocyanate (Upjohn Corp.).
[d]Trichlorofluoromethane.
[e]A mixture of Voranol RS 530 (Dow Chemical Co.) and I.

Table 14. Flammability Characteristics of Boron-Containing Foams

Foam No.	Density, pcf	% B	Burn rate,[a] in/min	Observations
1	2.96	1.4	10.1 ± 0.5	Shrinks
2	4.15	3.63	Immediate self-extinguishing	Melts
3	3.72	2.28	Self-extinguishing (7 sec)	Chars
4	16.9	1.4	Self-extinguishing (1-6 sec)	Chars, slight softening
5	3.54	1.26	Self-extinguishing (14 sec)	Chars, slight melting
6	3.36	1.3	Self-extinguishing (7-10 sec)	Chars, sags
7	3.69	0.89	7.3 ± 0.3	Surface char, shrinks

[a]Modified ASTM D 635-56T.

a hypothetical polymer which exhibits a simple unimolecular degradation process. Point A represents the region of initial decomposition, point B represents the point of maximum degradation.

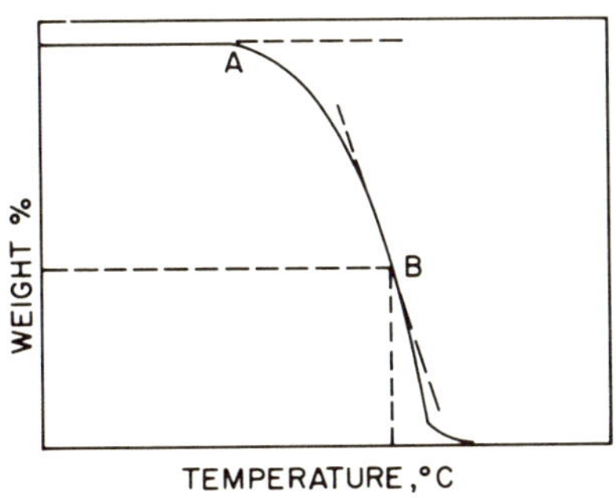

FIG. 25. Thermogram for a hypothetical polymer.

Fire retardants are screened in order to select a material having thermal characteristics similar to those illustrated in Fig. 23. A greater degree of stability can be obtained if two fire retardants are used: one fire retardant which changes state approximately 60°-75°C lower in temperature than point A, and a second fire retardant which changes state in the temperature range of point B. For polymers which degrade by means of a complex or autocatalytic process this technique may be used to select a balanced fire-retardant system using a combination of several retardants.

The choice of fire retardants is, of course, dependent on the nature of the polymer, the method of processing, the proposed service conditions, and economic considerations. Generally, reactive fire retardants are more stable than inert systems. Numerous investigations of materials exposed in actual fires have indicated that the nonreactive retardants may sublime out of the polymer when exposed to the hot gases which normally proceed the flame front. Figure 26 illustrates weight losses equivalent to the weight of fire retardant in

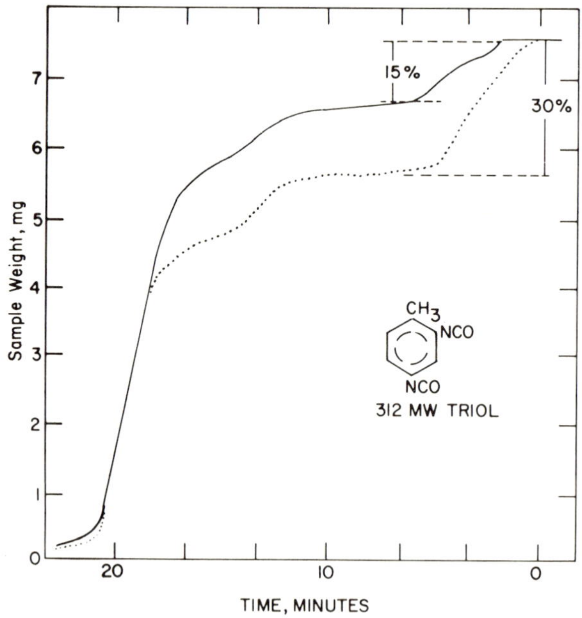

FIG. 26. Sublimation of fire retardant (vacuum).

urethane polymers when the nonreactive tris-2,3-dibromopropyl phosphate was used. Further studies were carried out in a vacuum of 10^{-7} Torr. Figure

27 shows a single sharp weight loss region when a non-fire-retarded urethane polymer was heated under dynamic conditions. The change in pressure resulted

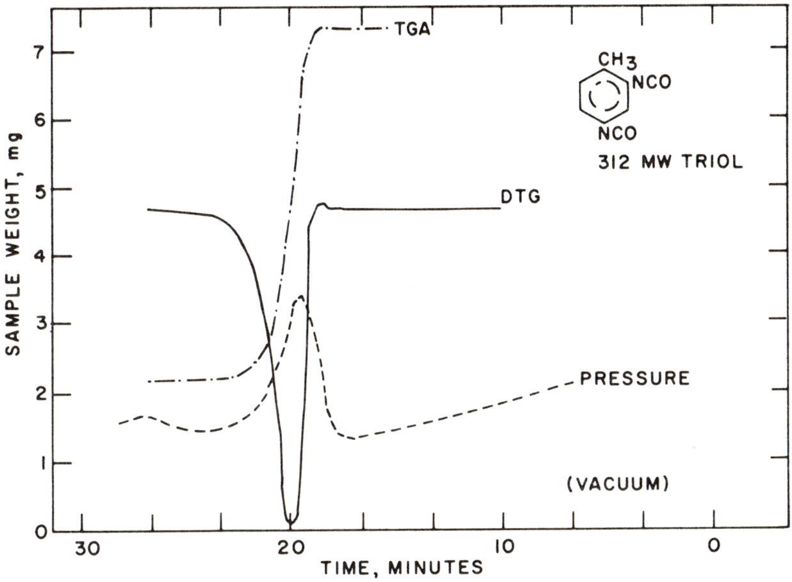

FIG. 27. Degradation of non-fire-retarded urethane polymer.

from the evolution of gaseous degradation products at the time of thermal decomposition of the polymer. Figure 28 shows a bimodal degradation process when the nonreactive fire-retardant tris-2,3-dibromopropyl phosphate was added to the model urethane. The initial weight loss (without an increase in pressure) resulted from sublimation of the nonreactive fire retardant. The material which was lost initially was collected and identified by normal chemical analysis procedures as the fire retardant.

In the process of selecting fire retardants to improve the flammability characteristics of a polymer or plastic, consideration must be given to total fire response and not just a single parameter such as flame propagation, smoke development, or fire endurance. The final product must be subjected to evaluation standards which approximate actual service life. For example, it is not normally possible to fire retard synthetic fibers when they are produced because of undesirable side effects during coloring, weaving, or sizing operations. Nonreactive fire retardants may be added as an aftertreatment with only minor

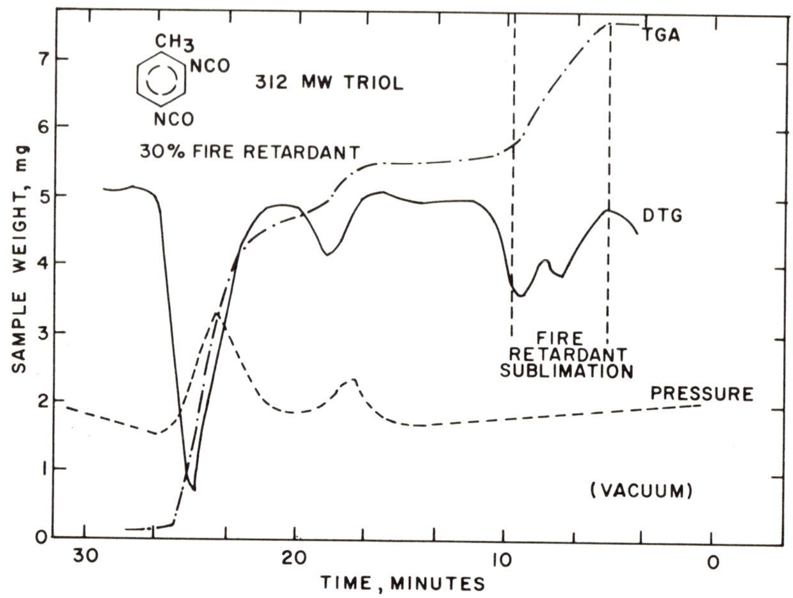

FIG. 28. Degradation of fire-retarded urethane polymer.

adverse effects to the process economics. Standardization tests usually are designed to rate the manufactured product. It is necessary to perform use-oriented tests in order to predict with the desired degree of confidence the performance of the finished product. Thus, in the case of fabrics, a given number of cycles of washing or dry cleaning may be used to evaluate the flammability characteristics after a preselected service life.

EVALUATION OF THE FLAMMABILITY CHARACTERISTICS OF POLYMERS

Flammability characteristics are those properties which describe the behavior of a material in actual fire exposure. Laboratory procedures which are used to classify these characteristics may be divided into the following seven categories:

1. Ease of ignition
2. Surface flame propagation
3. Heat contribution
4. Smoke development

5. Fire endurance
6. Products of degradation
7. Mechanism of degradation

Hilado [51] presented an excellent review of the testing procedures most commonly utilized in determining the flammability characteristics of plastics. Thus, rather than attempt to duplicate this survey the following discussion will review procedures which provide the maximum utility in screening fire retardants and fire-retarded plastics on a laboratory scale.

Ease of Ignition and Flame Propagation

A small-scale vertical test procedure has been developed which permits simultaneous evaluation of the ignition process and measurement of flame spread. A sample 10 by 2 by 1/8 in. is clamped in a vertical position with 1 in. calibration marks scribed to facilitate measurement of flame travel. The ignition source is a bunsen burner mounted on a track to permit positive positioning of the burner with respect to the sample. A single lens reflex camera with appropriate close-up attachments on a between-the-lens photometer is used to photograph the test specimen during combustion. Color infrared-sensitive film is employed to measure the response to fire exposure. This film will permit penetration of the flame front and smoke and then allow visual inspection of the substrate. Figure 29 shows the test specimen at the moment of flame contact. Figure 30 illustrates the point of sample ignition.

Heat Contribution. Several test methods have been developed which provide a measure of the amount of heat produced by a material during combustion. These tests usually do not provide a quantitative measure of the absolute amount of heat but rather a measure of the relative amount of heat generated. ASTM E-84 and ASTM E-162 test procedures are essentially designed to measure surface flammability, but both methods provide measures of heat contribution based on the temperature rise of the effluent gases.

Smoke Development. One of the many aspects that are of major concern when studying the flammability characteristics of materials is the analysis of the parameters governing the smoking tendency. As additives are incorporated into a polymer to retard its flammability, it is usually found that the quantity of smoke increases as flame propagation decreases. Most often the smoke emitted contains large amounts of carbon in the form of soot which readily obscures light.

In studies of the tendency of various fuels to smoke, Minchin (52) and

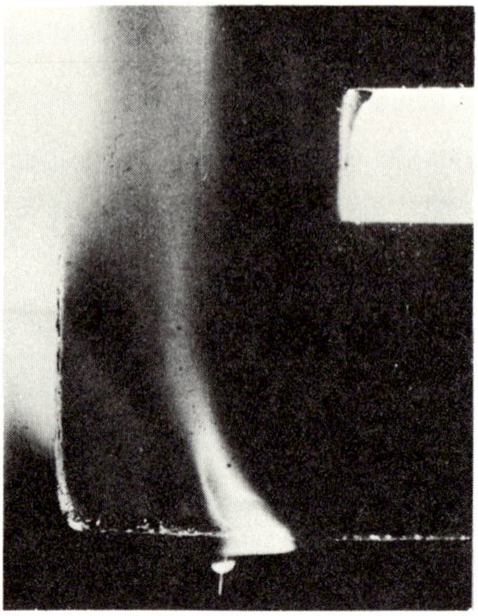

FIG. 29.

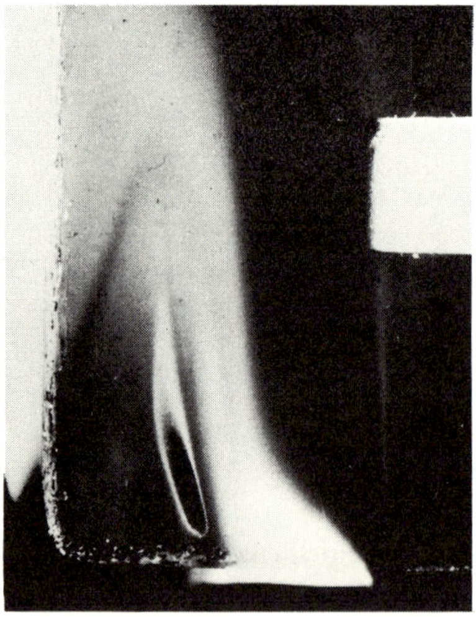

FIG. 30.

Clarke et al. [53] found that the tendency of paraffins to smoke increased with the molecular weight. On the other hand, for olefins, aromatic polymers, and compounds in the naphthalene series, the tendency to smoke was found to decrease with molecular weight. Figuie 31 summarizes the tendency of fuels to smoke.

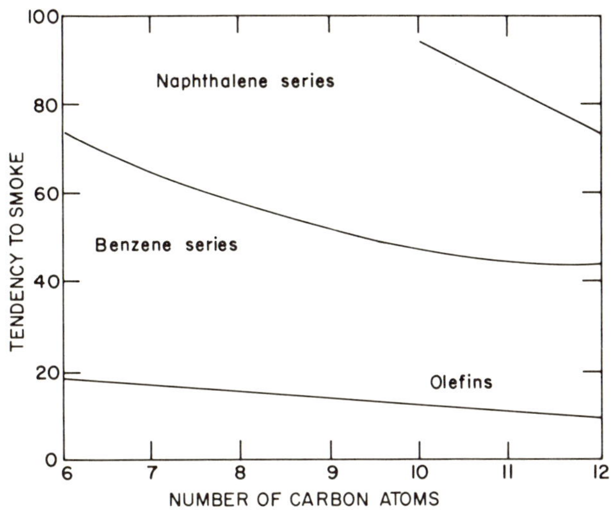

FIG. 31. Tendency of various fuels to smoke.

Michelson and Einhorn [54] reported on studies carried out to determine the effect of reactive and nonreactive flame retardants on the smoke development in urethane polymers. A series of model urethane foams were prepared containing varying amounts of o,o-diethyl-n,n-bis(2-hydroxylethyl) aminomethylphosphonate (a reactive fire retardant) and tris-2,3-dibromopropyl phosphate (a nonreactive fire retardant).

The smoke development properties of urethane foams were studied using a modified XP-2 smoke density chamber [55] illustrated in Fig. 32. The chamber measures 12 by 12 by 31 in. high, and it is constructed of anodized aluminum with stainless steel fittings. The door is cast aluminum with an inset fire-resistant plate glass panel to permit observation of the test specimens during burning.

A quantitative analysis of the smoke density is measured by the use of a Weston barrier layer photocell incorporated in the walls of the chamber 11 in. above the sample support. Normally, the time necessary to reach 100% light obscuration is used as a measure of the amount of smoke being produced. How-

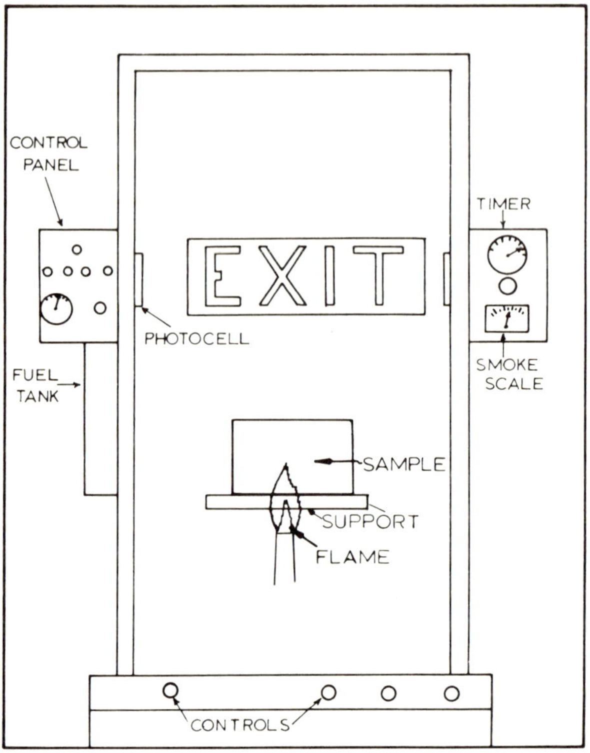

FIG. 32. Schematic drawing of XP-2 smoke chamber.

ever, since the approach to 100% light obscuration is asymptotic, it is very difficult to obtain times to 100% obscuration which are reproducible. Moreover, some samples can be burned completely without ever producing enough smoke to give 100% light obscuration. For these reasons, the time necessary to reach 70% light obscuration is taken as the arbitrary criterion for comparing the smoke developed in the burning of various foam samples. For samples of the same material, the time to 70% light obscuration can be measured within a standard deviation of 0.4 sec.

When a nonreactive fire retardant was added to the polymer system, the amount of smoke developed was found to increase as the concentration of the fire retardant was increased. This is shown in Fig. 33, where it was seen that the time to reach 70% light obscuration decreased with increasing fire-retardant concentration. A similar reduction in the time to reach 70% light obscuration

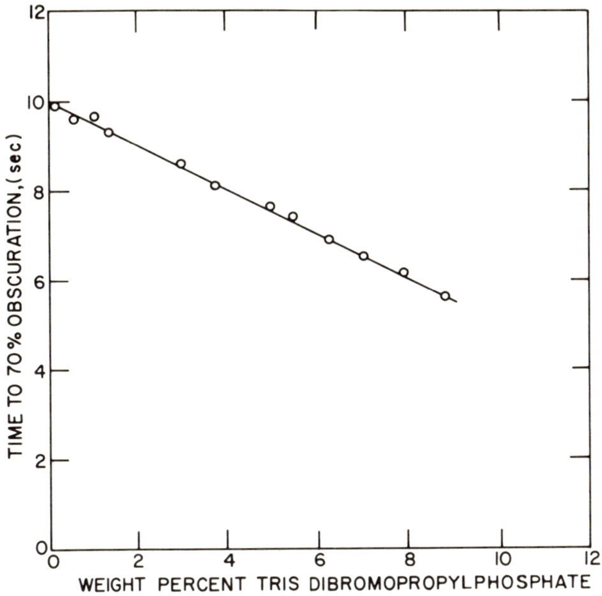

FIG. 33. The effect of a nonreactive fire retardant on smoke development.

was shown in Fig. 34 for the reactive fire retardant. For both systems, there seems to be a limiting value for the time required to reach 70% light obscuration as the concentration of the fire retardant was increased. In the case of the nonreactive fire retardant, this value was about 5 sec; whereas, for the reactive fire retardant, this value was approximately 3 sec.

When the horizontal ASTM D-1692 test was run for flame spread, a definite correlation was found between the distance that the flame travels before extinguishing itself and the amount of smoke developed during burning as Fig. 35 shows for a nonreactive and a reactive fire retardant. This figure shows that as materials become more difficult to burn, they produce more smoke and, in the region where the samples are considered self-extinguishing, there is a straight-line relationship between the obscuration of light caused by smoke and the flame spread.

The temperatures at which thermal decomposition occurs were determined by differential thermal analysis for polymers containing the reactive and nonreactive fire retardants. The results are shown in Figs. 36 and 37. In an effort to determine if a correlation might exist between the temperatures at which thermal decomposition occurs and the amount of smoke a polymer produces when it is burned, it was assumed that a relationship between these two quantities can be expressed as

$$\theta = mT + b$$

where θ is the time to 70% obscuration, and T is the temperature at which thermal decomposition occurs; m and b are arbitrary constants and they must

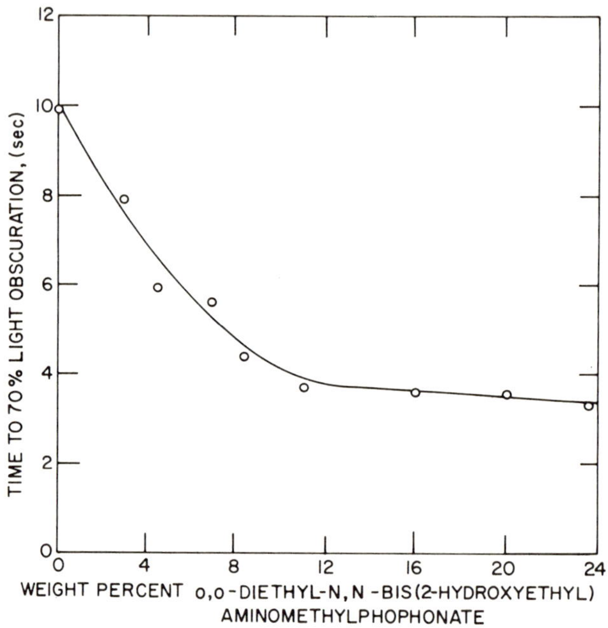

FIG. 34. The effect of a reactive fire retardant on smoke development.

be determined from two data points. The above equation is the simplest expression that can be used to normalize the data. Using the two end data points for both the nonreactive and the reactive fire retardants, the following equations were found by which smoke obscuration can be related to the temperature at which thermal decomposition occurs.

Nonreactive: $\theta = 0.0434T - 9.1$

Reactive: $\theta = 0.091T - 24.7$

Figures 38 and 39 show the comparison that results between the smoke obscuration predicted by the above equation and the measured values for

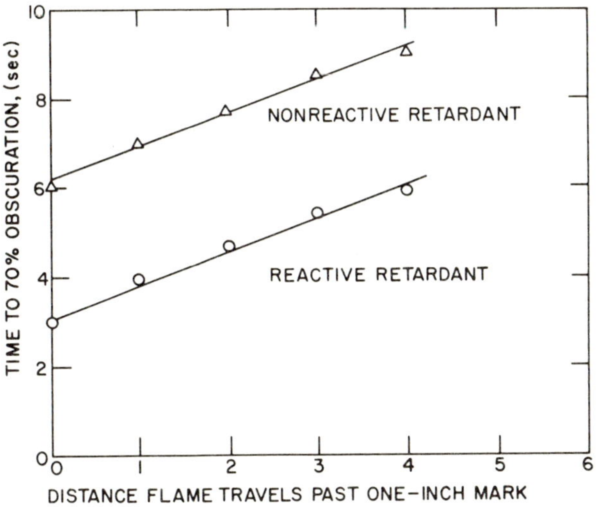

FIG. 35. Smoke obscuration compared to flame travel.

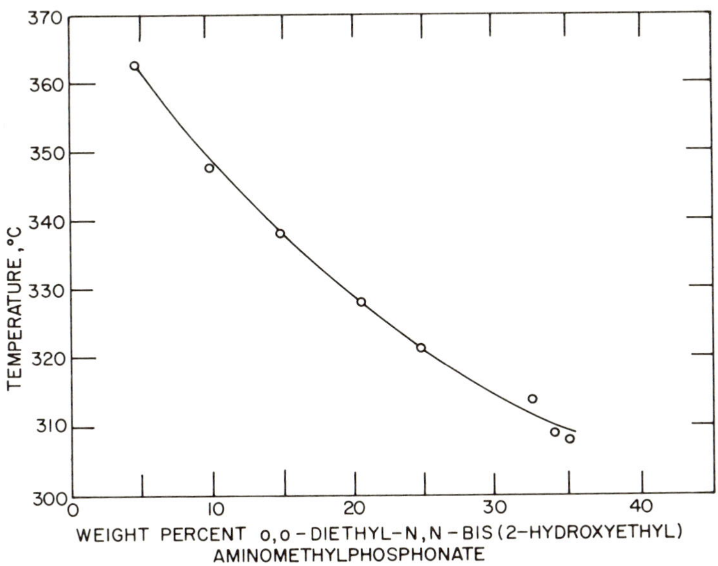

FIG. 36. Thermal decomposition as determined by DTA analysis for a polymer with a reactive fire retardant.

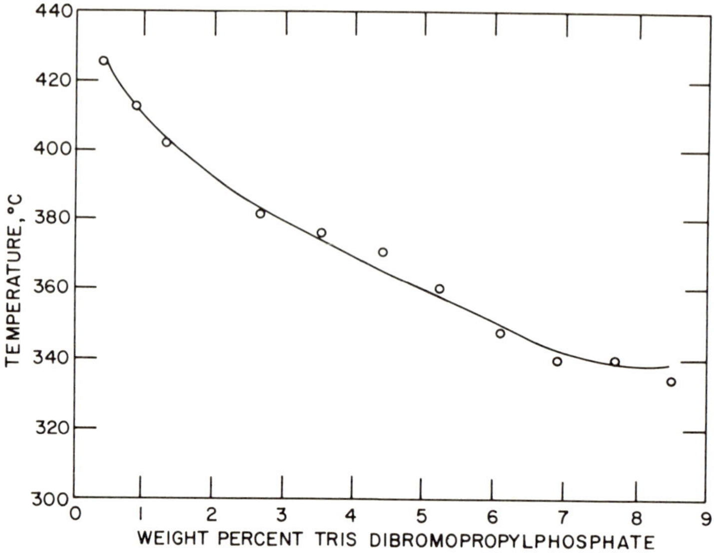

FIG. 37. Thermal decomposition as determined by DTA analysis for a polymer with a nonreactive fire retardant.

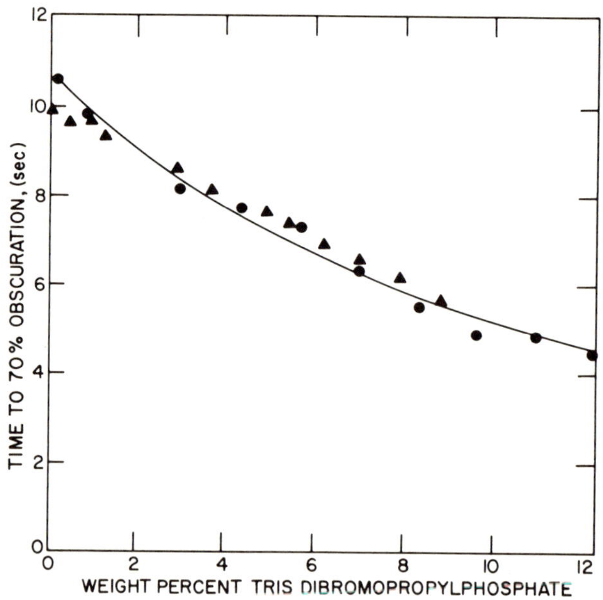

FIG. 38. Comparison of DTA data with data on smoke development for nonreactive fire retardant: (●) times to 70% obscuration and (▲) normalized data from DTA.

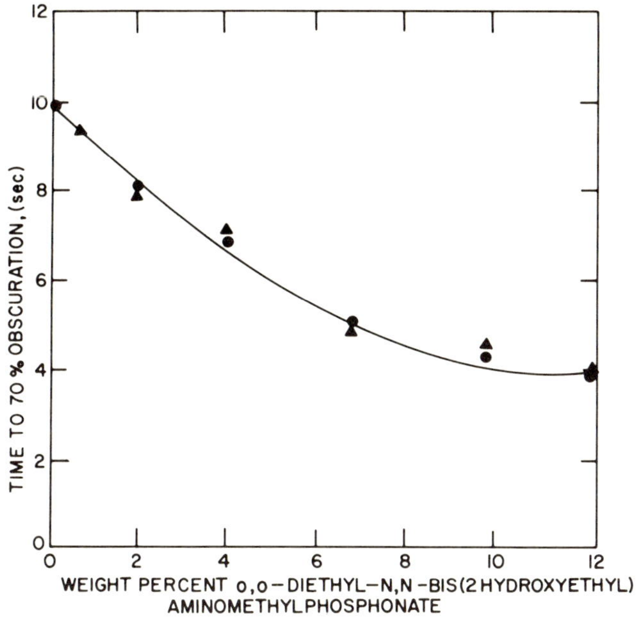

FIG. 39. Comparison of DTA data with data on smoke development for reactive fire retardant.

smoke obscuration for the nonreactive fire retardant and the reactive fire retardant. Both figures show that there was good agreement between predicted values and measured values. This agreement certainly indicated that a possible relationship exists between the temperatures at which thermal decomposition occurs and the amount of smoke formed during burning.

Two major deficiencies exist in the testing procedures used to evaluate smoke development. One of these deficiencies relates to a lack of measurement of the quantity of material that must burn to produce a given quantity of smoke. Einhorn and co-workers [55] reported on this parameter when studying low-density fire-retardant materials. The sample was connected to the balance arm of a microbalance and burned in the XP-2 smoke density chamber. Figure 40 shows results obtained when urethane, polyimide, and polyisocyanurate foams were burned in the smoke chamber. The second deficiency arises when an attempt is made to correlate the results of smoke development with the physiological response. When materials such as wood burn, little smoke is developed; thus, if this material is evaluated in the smoke density chamber,

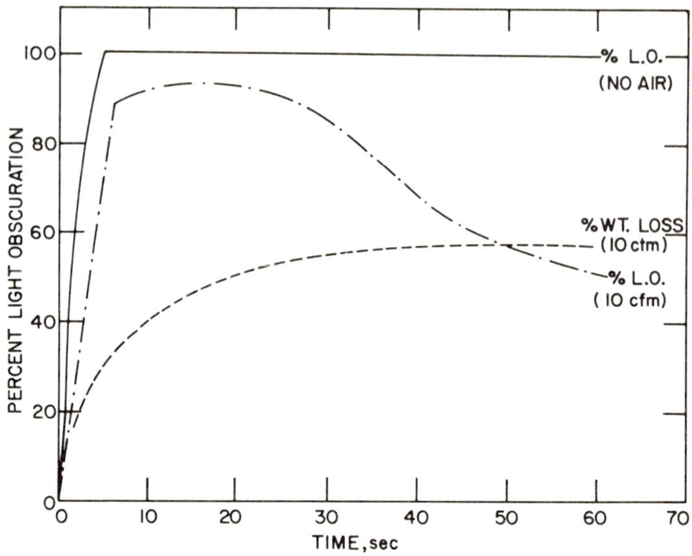

FIG. 40. Thermal decomposition of 5-1 foam – smoke development. Here L. O. stands for light obscuration.

the photometer registers little smoke. However, a human placed in such a chamber (or exposed to fire under similar conditions) would not be able to see because some of the degradation products cause the eye to tear, thus reducing vision.

Fire Endurance. Fire endurance is a measure of resistance to "burn-through" or erosion of a material during fire exposure. Figures 41 and 42 show schematic diagrams of the horizontal and vertical assemblies for the Bureau of Mines fire endurance apparatus. Einhorn [56] reported on the fire endurance characteristics of cellular plastics using fire retardants and inert extenders. Tables 15 and 16 show the results obtained using the horizontal and vertical procedures.

Products of Degradation. Many investigators use large samples and thus are able to determine the products of degradation by ordinary chemical methods of organic analysis. There are several major difficulties associated with such procedures. Organic chemical analysis is time consuming in view of the complex nature of the degradation products. Additional complications arise from the rise of large samples.

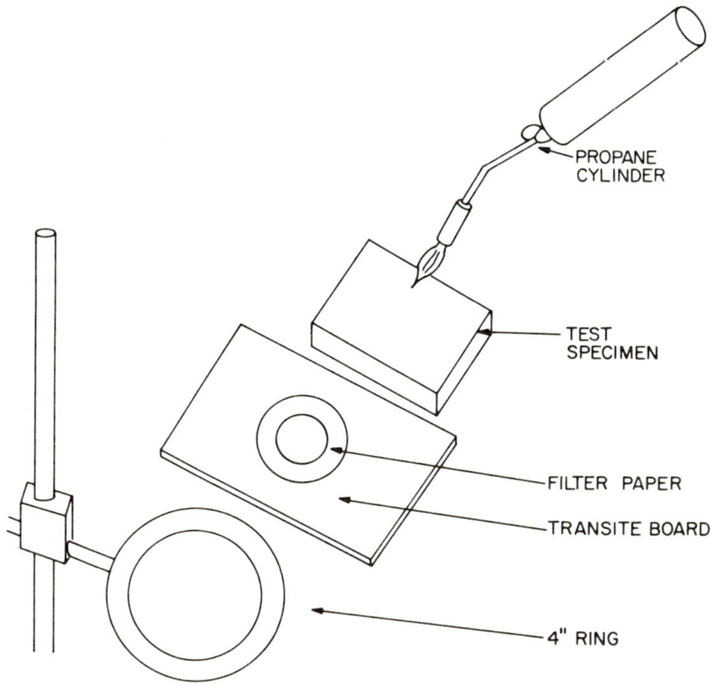

FIG. 41. Schematic drawing of Bureau of Mines apparatus (horizontal test method).

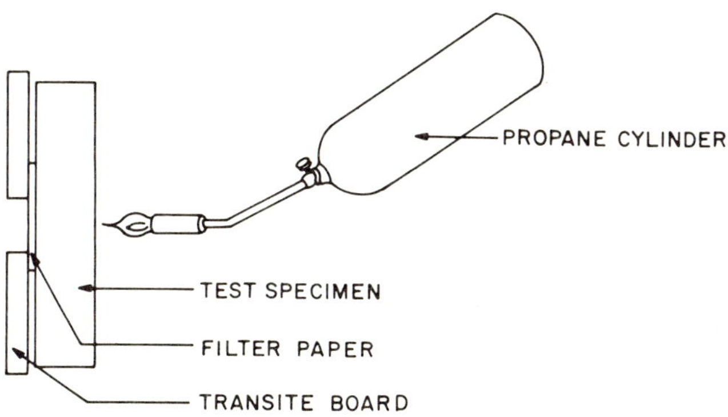

FIG. 42. Schematic drawing of Bureau of Mines apparatus (vertical view).

Table 15. Bureau of Mines Fire Endurance Test (Horizontal Method)

Sample description	Endurance time
Cellular plastics	
Polystyrene bead board (non-S.E.)	7 sec
Expanded polystyrene (non-S.E.)	7 sec
Expanded polystyrene (S.E.)	9 sec
Urethane foam (non-S.E.)[a]	20 sec
Epoxy foam (S.E.)	1 min 20 sec
Urethane foam (S.E.)	19 min 20 sec
Cellular composites	
Styrene bead foam/glass pellet composite	9 sec
Styrene bead foam/pelletized glass microballoons composite	11 sec
Urethane foam (non-S.E.)[a]/mica composite	20 sec
Urethane foam (non-S.E.)/glass flake composite	20 sec
Urethane foam (non-S.E.)[a]/vermicrilite composite	5 min 22 sec
Urethane foam (non-S.E.)[a]/glass flake composite	39 sec
Urethane foam (non-S.E.)[a]/glass pellet composite	10 min 40 sec
Urethane foam (non-S.E.)[a]/pelletized glass microballoons composite	10 min 48 sec
Urethane foam (S.E.)[b]/glass flake composite	3 hr +
Urethane foam (S.E.)[b]/glass pellet composite	3 hr +
Urethane foam (S.E.)[b]/pelletized glass microballoons composite	3 hr +
Miscellaneous control samples	
Fiber glass board (12.0 pcf)	7 min 45 sec
White pine (25.2 pcf)	53 min 15 sec

[a]Urethane foam (TDI, sorbitol adduct, F-11 system).
[b]Urethane foam (polymeric isocyanate, chlorinated polyester polyol, F-11 system).

Table 16. Modified Bureau of Mines Fire Endurance Test (Vertical Method)

Sample description	Endurance time
Cellular plastics	
Polystyrene bead board	9 sec
Expanded polystyrene (non-S.E.)	10 sec
Expanded polystyrene (S.E.)	12 sec
Urethane foam (non-S.E.)[a]	38 sec
Epoxy foam (S.E.)	5 min 44 sec
Urethane foam (S.E.)[b]	14 min 2 sec
Cellular composites	
Styrene bead foam/glass pellets composite	11 sec
Styrene bead foam/pelletized glass microballoons composite	12 sec
Urethane foam (non-S.E.)[a]/mica composite	1 min 40 sec
Urethane foam (non-S.E.)[a]/glass flake composite	1 min 44 sec
Urethane foam (non-S.E.)[a]/vermicalite composite	1 min 53 sec
Urethane foam (non-S.E.)[a]/glass flake composite	1 hr 10 min
Urethane foma (non-S.E.)[a]/pelletized glass microballoons composite	1 hr 15 min
Urethane foam (S.E.)[b]/glass flake composite	2 hr 5 min
Urethane foam (S.E.)[b]/glass pellet composite	3 hr + (test discontinued)
Urethane foam (S.E.)[b]/pelletized glass microballoons composite	3 hr + (test discontinued)
Miscellaneous control samples	
Fiber glass board (12.0 pcf)	10 min 10 sec
White pine (25.2 pcf)	1 hr 30 min

[a]Urethane foam (TDI, sorbitol adduct, F-11 system).

[b]Urethane foam (polymeric isocyanate, chlorinated polyester polyol, F-11 system).

Decomposition of most organic polymers is an endothermic reaction. Polymers are insulators; thus, because of their poor thermal conductivity the temperature profile is not uniform when a large volume of polymer is heated. Another difficulty arises from the fact that diffusion of the degradation products from the inside of a large sample may be retarded, thus giving rise to secondary reactions. A uniform temperature is essential to the correct interpretation of rates of degradation of the polymer and the mechanism of formation of the degradation products as functions of temperature.

Einhorn et al. [55] conducted experiments to determine the degradation products of model urethane polymers under pyrolysis conditions. The thermal degradation characteristics of the polymers were determined using simultaneous DTA, TGA, and DTG techniques. Figure 43 shows a typical thermogram

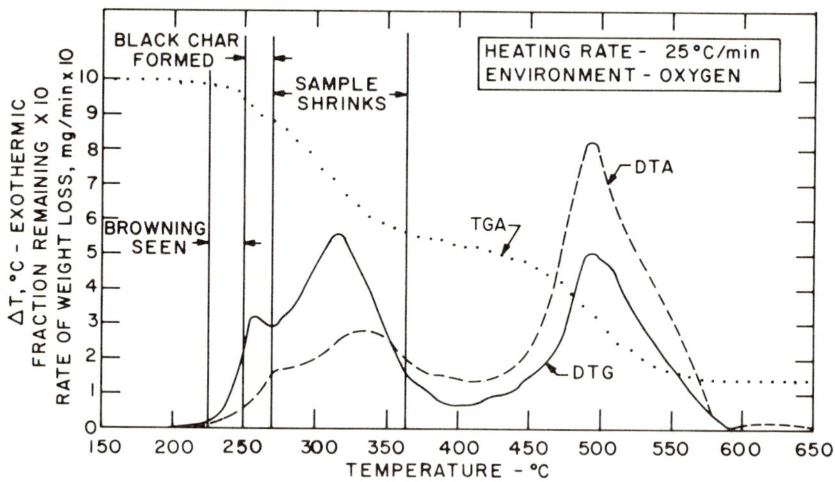

FIG. 43. Visual observation of 5-1 foam decomposition correlated to the thermal analyses.

covering the complete degradation process. The temperature ranges for individual degradation reactions were determined and isothermal pyrolysis experiments carried out to identify the products of combustion. Table 17 shows the light gases identified using a thermal conductivity detector. Figure 44 shows the many fragments obtained when similar studies were carried out using a flame ionization detector. Additional information can be obtained using a mass spectograph or infrared techniques.

Table 17. Pyrolysis Products – Urethane Foam (Light Gases)

Pyrolysis temp, 180°C	Pyrolysis temp, 300°C	Pyrolysis temp, 350°C	Pyrolysis temp, 600°C
0_2(trace)	0_2	0_2	H_2
			0_2
N_2(trace)	N_2	N_2	N_2
		CO	CH_4
CO_2(trace	CO_2	CO_2	CO
			N_20
			CO_2

Mechanism of Degradation. Michelson and Einhorn [57] used several of the methods described in the section entitled "Theoretical Considerations of Thermal Degradation" to determine the kinetic parameters of degradation. A urethane polymer was prepared by reacting pure 2,4-toluene diisocyanate with a propoxylated trimethylol propane (MW 312). No catalyst was used in this reaction. The polymer was cast into small slabs and cured at temperatures below 80°C to insure that any biuret allophanate formed as a side reaction would be minimal.

A typical thermogram for the decomposition of the materials is shown in Fig. 45. The maximum point method for determining the kinetic parameters was applied to the data and the results are presented in Table 18. These results indicate that this method gives widely varying answers. Of particular interest are the activation energies resulting from data obtained at a heating rate of 10°C/min. These results show a variation of 9000 cal/g mole in the activation energy for the decomposition of the polymer. Such a large variation supports the premise of Flynn and Wall about the difficulty in determining reliable activation energies using this method.

The Freeman and Carroll method was also used to analyze the data, and a summary of the results of this analysis is shown in Table 19. Here, it is shown that much better agreement is obtained between the activation energies calculated for heating rates of 10°C/min. The variation in the activation energies from the highest value and the lowest value amounts to only 10%. Also, the results obtained at a heating rate of 10°C/min are in much better agreement with the results obtained at a heating rate of 4°C/min.

The results obtained by the ratio method are presented in Table 20. Except

Table 18. Determination of Kinetic Parameters (Maximum Point Method)

Heating rate, °C/min	Activation energy, cal/g mole	Reaction order	Frequency factor, min^{-1}
4	37,600	0.46	6.3×10^{13}
10	36,000	0.59	2.6×10^{13}
10	31,000	0.47	1.6×10^{11}
10[a]	27,000	0.38	5.6×10^{9}

[a]Sample weight is 4.0 mg.

Table 19. Determination of Kinetic Parameters (Freeman-Carroll Method)

Heating rate, °C/min	Activation energy, cal/g mole	Reaction order	Frequency factor, min^{-1}
4	34,600	0.50	4.58×10^{12}
10	33,400	0.20	2.18×10^{12}
10	30,300	0.50	1.29×10^{11}
10[a]	32,700	0.45	1.23×10^{12}

[a]Sample weight is 4.0 mg.

Table 20. Determination of Kinetic Parameters (the Ratio Method)

Heating rate, °C/min	Activation energy, cal/g mole	Reaction order	Frequency factor, min^{-1}
4	33,200	0.46	1.32×10^{12}
10	33,200	0.55	1.51×10^{12}
10	30,500	0.61	1.55×10^{11}
10[a]	33,200	0.49	1.91×10^{12}

[a]Sample weight is 4.0 mg.

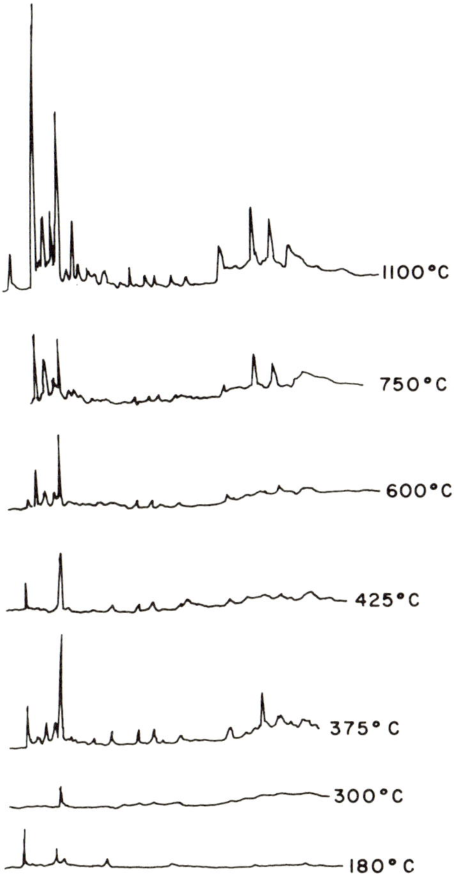

FIG. 44. Thermal decomposition of 5-1 10AOB foam.

for the one run at 10°C/min, there is excellent agreement between the activation energies obtained at heating rates of 4°C/min and at heating rates of 10°C/min.

The data presented in Fig. 46 represent a ratio of γ_i/γ_j of 1.5 to 1 and Fig. 47 represents a ratio of γ_i/γ_j of 3 to 1. By using high ratios such as 3 to 1, errors in measuring the temperature or the decomposition rate will produce less scatter in the data. Lower ratios such as 1.5 to 1 are more sensitive to these errors. For example, if the difference between temperatures T_j and T_i is large, then any error in measuring these temperatures will produce less

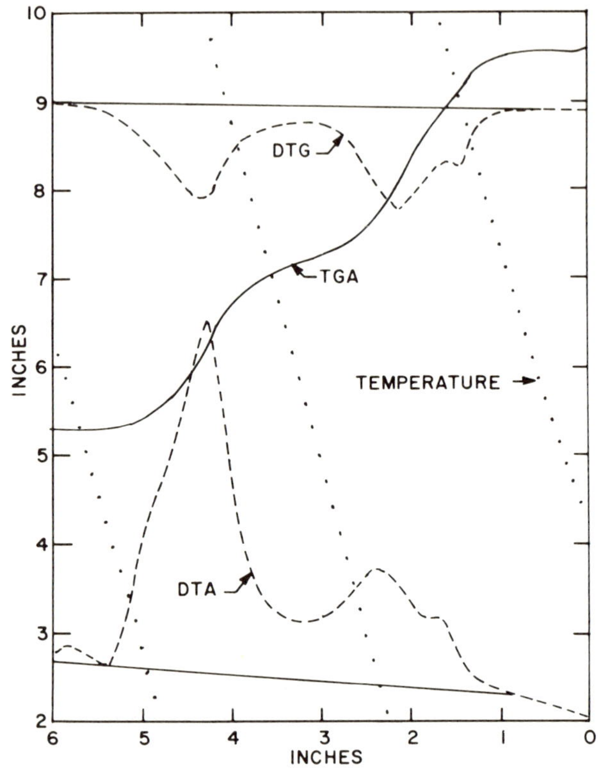

FIG. 45. Typical curves of data.

scattering in the data. Lower ratios are more sensitive to the same errors because the difference between T_j and T_i gets smaller as the ratio gets smaller. However, as can be seen in Fig. 46, very little scatter exists indicating a reasonable accuracy in measuring those variables from which the kinetic parameters are determined.

For both the Freeman and Carroll method and the ratio method of analyzing kinetic data, only the portion of the thermogram from 90% undecomposed material down to 10% undecomposed material was examined. This was done because either inaccuracies appear in measuring the thermal decomposition at the extremities of the thermogram or the power-model kinetic equation does not apply over the entire decomposition range. With the ratio method, it is possible to investigate the extremities of the thermogram and determine what

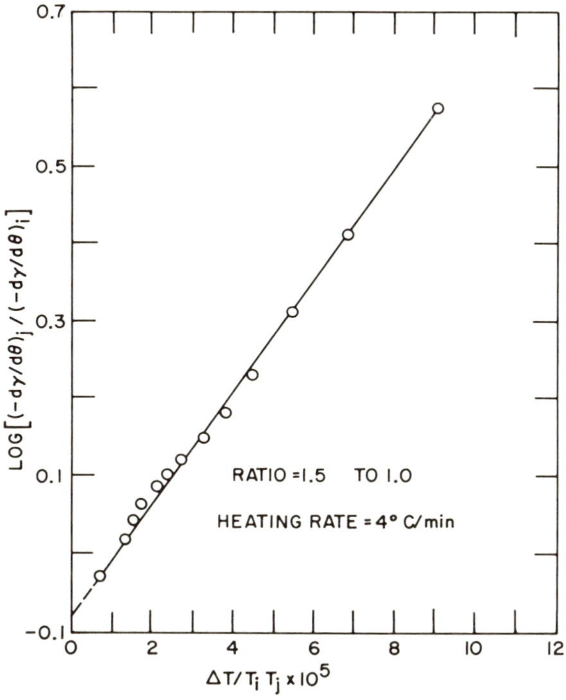

FIG. 46. The ratio method supplied to decomposition at a low heating rate.

deviations are occurring. If Eq. (31) is rearranged into the following form

$$\frac{\log (-d\gamma/dt)_j/(-d\gamma/dt)_i}{\log (\gamma_i/\gamma_j)} = \frac{E}{2.303R \log (\gamma_i/\gamma_j)} \frac{T_j - T_i}{T_iT_j} \tag{31}$$

a straight line should be obtained regardless of what ratio is used. It is permissable to divide by the log (γ_i/γ_j) because this quantity is a known constant and not a variable as is the case when dividing by d log γ.

An end point analysis is made for one of the samples heated at a rate of 10°C/min, and the results are shown in Fig. 48. The circle points cover a decomposition range from 90% undecomposed material down to 10% undecomposed material. The ratio γ_i/γ_j was varied from 9 to 1 down to 1.5 to 1. When the point corresponding to 95% of the material undecomposed is compared to other values of γ, the line represented by the designation of an open triangle is obtained. This line shows that the earlier stages of decomposition

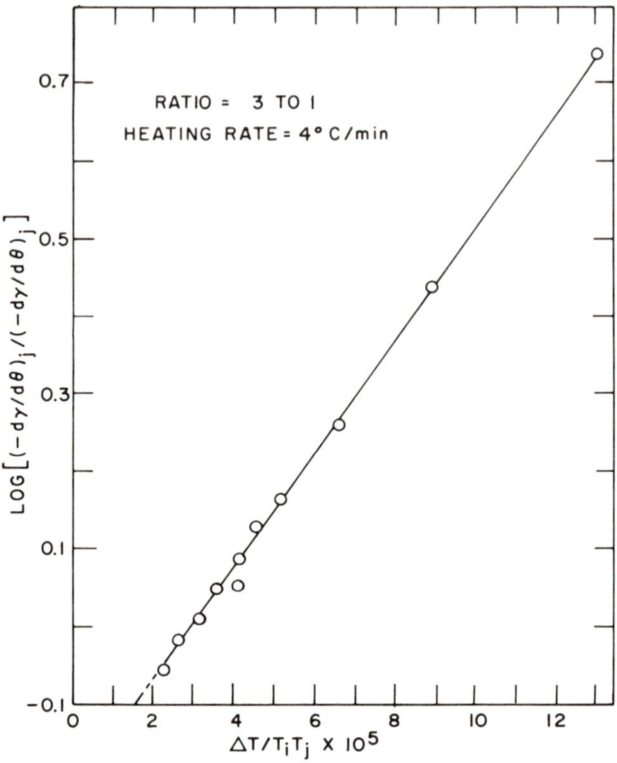

FIG. 47. The ratio method applied to decomposition at a low heating rate.

definitely do not obey the same mechanism of decomposition as do the later stages. When the point corresponding to only 5% of the material remaining undecomposed is compared to other values of γ, a line indicated by the filled triangle is found to lie parallel to the original line. This indicates that the same activation energy governs the decomposition in the lower region, but the reaction order is slightly changes. The excellent fit of the data to the power-model kinetic equation over the major portion of the decomposition reaction indicates that this model equation is well suited for explaining the decomposition kinetics of a simple urethane polymer.

The ratio method has proved to be a simple and accurate tool for determining kinetic constants. Just as in the method of Freeman and Carroll, the entire thermogram can be used in determining the kinetic parameters. However, in the ratio method, it is not necessary to determine the slopes of the logarithm

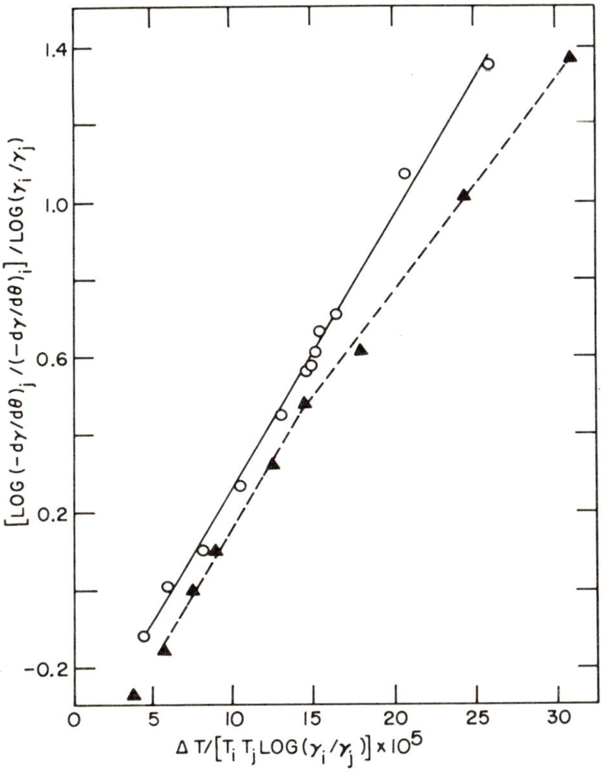

FIG. 48. End point analysis of a thermogram: (○)γ_{ij} = 0.1 to 0.9, (△)γ_j = 0.93, and (▲)γ_i = 0.05.

of γ and of the logarithm of $(-d\gamma/dt)$ as is required by the method of Freeman and Carroll. Although the maximum point method permits the rapid analysis of a thermogram, it has proved to be far less accurate than the ratio method or the method of Freeman and Carroll. Even though the ratio method is not as fast as the maximum point method for examining thermograms, it is far more accurate and it is readily adaptable to the use of computers in the analyses of thermograms.

The entire thermogram should always be analyzed when determining kinetic parameters to insure that there is either no change in the reaction mechanism or that no competing reactions are occurring. If it is suspected that the decomposition reaction is not a simple one, then separate portions of the thermogram can be examined by selecting ratios of γ_i/γ_j that give enough points to determine a reliable correlation covering the desired portion of the thermogram.

Although the main purpose of this investigation was to establish the validity to the ratio method for analyzing thermogravimetric data, it was found that for the urethane studied the kinetic equation is

$$(-d\gamma/dt) = 1.58 \times 10^{12} [\exp(-33{,}200/RT)] \gamma^{1/2} \min^{-1} \tag{32}$$

The one result at 10°C/min where the activation energy is low is not included in determining the average value of the kinetic constants.

SUMMARY AND CONCLUSIONS

The combustion of polymeric materials is a very complex process. Considerable research and development efforts are being expended to improve the flammability characteristics of these materials. The development of a fundamental understanding of the nature of the mechanism of thermal degradation and combustion of polymers must be gained if the desired flammability characteristics are to be modified to meet new and more stringent regulations.

REFERENCES

[1] E. A. Boettner, *Proc. Polymer Conf. Ser., Univ. of Detroit, June 1969.*
[2] J. DiPietro, *Proc. Polymer Conf. Ser., Univ. of Detroit, June 1969.*
[3] J. D. Seader, *Proc. Polymer Conf. Ser., Univ. of Detroit, June 1969.*
[4] C. P. Fenimore and F. J. Martin, *Proc. Polymer Conf. Ser., Wayne State Univ., June 1966.*
[5] E. A. Dickert and G. C. Toone, *Mod. Plastics,* **42**, 5, 197 (1965).
[6] L. Straehl, "Textile Flammability and Consumer Safety," Gottlieb Duttweiler Inst. for Economic and Social Studies Rüschlikon Zurich, 1969.
[7] G. F. Crikelair, "Textile Flammability and Consumer Safety," Gottlieb Duttweiler Inst. for Economics and Social Studies Rüschlikon Zurich, 1969.
[8] I. N. Einhorn and R. W. Mickelson, *Proc. Div. of Organic Coatings and Plastics, San Francisco, Calif., March 1968, Am. Chem. Soc.,* **28**(1), 291-310 (1968).
[9] G. M. Lukaszewski and J. P. Redfern, *Lab. Pract.,* **10**, 721 (1961).
[10] K. Akita and M. J. Kase, *J. Polymer Sci.,* **54**, W-3 (1961).

[11] J. H. Flynn and L. A. Wall, *Polymer Letters,* **4**, 323-328 (1966).
[12] H. L. Friedman, *J. Macromol. Sci.,* **59** (1967).
[13] R. W. Mickelson and I. N. Einhorn, *Thermochem. Acta* **1**, 147-158 (1970).
[14] I. N. Einhorn, *Proc. Polymer Conf. Ser., Univ. of Detroit, June 1969.*
[15] G. B. Skinner, "Survey of Chemical Aspects of Flame Extinguishment," ASO Tech. Rept. 61-408, 1961.
[16] G. Lash and H. G. Wagner, *Eighth Intern. Symposium on Combustion, Pasadena, 1960;* abstracts of papers pp. 55-6.
[17] H. Piechota, *J. Cell Plast.* **1**, 186-199 (1) (Jan. 1965).
[18] J. N. Tilley, *Proc. Polymer Conf. Ser. Wayne State Univ., June 1966.*
[19] R. Friedman and J. B. Levy, "Survey of Fundamental Knowledge of Mechanisms of Action of Flame Extinguishing Agents," WADC-TR-56-568, Jan. 1967.
[20] P. Z. Li *et al., Plasticheskie Massy,* (5) 15-19 (1963).
[21] Al'shits *et al., Zhur. Priklad. Khim.* **34**, 468-9 (1961).
[22] United States Rubber Co., British Pat. 924,323 (April 24, 1963).
[23] Dominion Rubber Co. Ltd., Can. Pat. 681, 805 (March 10, 1964).
[24] Hooker Chemical Corp., U. S. Pat. 3,007,958 (Nov. 7, 1961).
[25] Rutgerswerke A. G., U. S. Pat. 3, 113,933 (Dec. 10, 1963).
[26] Michigan Chemical Corp. British Pat. 988,304 (April 7, 1945).
[27] Deutsche Akademic der Wissenchaften zu Berlin, French Pat. 1,294,986 (April 24, 1962).
[28] Michigan Chemical Corp., Belgian Pat. 655,798 (Nov. 16, 1964).
[29] R. C. Nametz and R. J. Nulph, *Proc. Ann. Tech. Congr. SPI, Reinforced Plastics Div., 20th Chicago, 1965,* Sec. 11-C.
[30] R. C. Nametz, J. DiPietro, and I. N. Einhorn, *Proc. Div. of Organic Coating and Plastics, Am. Chem. Soc.* **28**, 204-224 (1) (April 1968).
[31] Hooker Chemical Corp., Test Method 15.
[32] D. W. Mitchell and E. M. Murphy, *Proc. Conf. on Foamed Plastics NAS-NRC, April 1963,* Publ. PB 181576.
[33] C. J. Hilado, *Flammability Handbook For Plastic,* Technomic Publ. Co., Stamford, Conn., 1969.
[34] N. J. Read and E. G. Heighway-Bury, *J. Soc. Dyers Colourists,* **74**, 823-30 (1958).
[35] R. W. Little, *Flameproofing Textile Fabrics,* Chem. Soc. Monogr. Ser. No. 104, Reinhold, New York, 1947.
[36] Dow Chemical Co., Brit. Pat. 924,945 (1960).
[37] C. T. Pumpelly, in *Bromine and Its Compounds* (Z. E. Jolles, ed.), Chap. 6, Academic Press, New York, 1966.
[38] W. G. Schmidt, *Trans. J. Plastics Inst.,* 247 (Dec. 1965).

[39] G. L. Drake, Jr., in *Fire-Resistant Textiles, Encyclopedia of Chemical Technology* (R. E. Kirk and D. F. Othmer, eds.), Vol. 9, Wiley, New York, 1966.
[40] J. Pitts, P. H. Scott, and D. S. Powell, *J. Cell. Plast.*, **6** (1) (Jan./Feb. 1970).
[41] I. N. Einhorn, N. Sonpal, and M. Kanakia, *J. Appli. Polymer Sci.* (in press).
[42] H. W. Eickner, *Fire Retardant Treated Wood*, American Soc. Testing Materials, Seattle, Washington, 1965.
[43] J. D. McCluer, U. S. Pat. 2,948,641 (1960) to Thermoid Co.
[44] J. Knaggs, Brit. Pat. 575,028 (1946).
[45] B. Rudner and M. S. Moores, U. S. Pat. 3,042,636 (1962) to Koppers Co. Inc.
[46] R. F. Schwenker, Jr., and L. R. Beck, Jr., *J. Polymer Sci.*, (2), 331 (1963).
[47] F. J. Kilzer and A. Broido, *Pyrodynamics*, **2**, 151 (1961)
[48] W. G. Woods, I. S. Bengelsdorf, and D. Laruccia, *Proc. Polymer Conf. Ser., Wayne State University, June 1966.*
[49] W. G. Woods, I. S. Bengelsdorf, and D. Laruccia, U. S. Pat. 3,189,565 (1965) to United State Borax and Chemical Corp.
[50] I. N. Einhorn, *Proc. Polymer Conf. Ser., University of Utah, 1970.*
[51] C. J. Hilado, *Flammability Handbook For Plastics,* Technomic Publ. Co., Stamford, Conn., 1969.
[52] S. T. Minchin, *J. Inst. Fuel,* **22**, 299 (1949).
[53] A. E. Clarke, T. G. Hunter, and F. H. Garner, *J. Inst. Petrol,* **32**, 627 (1946).
[54] R. W. Mickelson and I. N. Einhorn, *Org. Coatings Plastics Chem.,* **28**, 272-281 (1) (April 1968).
[55] I. N. Einhorn, J. D. Seader, and N. Sonpal, *Proc. Polymer Conf. Ser., Univ. of Utah, 1970.*
[56] I. N. Einhorn, *Proc. Polymer Conf. Ser., Univ. of Detroit, June 1969.*
[57] R. W. Mickelson and I. N. Einhorn, *Thermochem. Acta* **1**, 147-158 (1970).

Recent Advances in Photo-Cross-Linkable Polymers

G. A. DELZENNE

Photochemical Research Laboratories
Agfa-Gevaert N. V.
Mortsel, Belgium

ABSTRACT

A short survey is given of recent advances in the field of photo-cross-linkable polymers. New synthetic approaches to known photosensitive polymers as well as new polymers are described. The photosensitization mechanism of cinnamic derivatives is examined in detail and the most striking arguments in favor of a triplet-triplet energy transfer process are discussed. The photosensitization mechanism of azido derivatives has also been taken into consideration. New sensitizers are described and fluorescence quenching experiments with anthracene are mentioned. Monoradical photo-cross-linking systems based on polymers carrying lateral unsaturation, photolabile groups, or photoreducible dye fragments are discussed. Finally, some remarks about future trends on the photo-cross-linking field are formulated.

1. HISTORICAL

Very few photochemical reactions have led to such important technical applications and can avail themselves of so ancient a history as the insolubilization reactions of photosensitive polymers.

It is customary to start a review devoted to photosensitive polymers by a short historical review going back to the Babylonian and Egyptian epochs [1]. A short mention of those almost prehistorical applications ensures the reader's attention without fail.

Egyptian mummies were wrapped with linen cloths whose photosensitivity was obtained by dipping in a solution of oil of lavender containing a high-molecular-weight bituminous material called Syrian asphalt or bitumen of

Judea. By exposure to sunlight the substance becomes cross-linked. The same bitumen was also used as a calking tar in Noah's ark and in the rush basket of the infant Moses, as told in the Bible. After these periods one has to await the work of Niepce to discover again the hidden potentialities of those techniques. Niepce, an amateur Egyptologist, produced the first permanent photograph in 1822 with a glass plate coated with Syrian asphalt. After exposure to bright sunlight the plate was developed in turpentine and the cross-linked exposed areas remained on the plate, which served as a lithographic surface for greasy inks [2].

However, the real rise of modern photosensitive polymers started in 1930 with the pioneer work of A. Murray who showed that unsaturated ketones could be photo-cross-linked to form resists [3]. A great number of modern technological developments would be unrealizable without the help of these derivatives.

Photosensitive polymers are the active ingredients of the photosensitive resists which may be cast into films on various substrates and rendered insoluble by exposure to actinic light. They find applications in the manufacture of microelectronic components, miniature and integrated circuits, photoengraving, and precision chemical milling.

Photoengraving is one of the basic techniques in processing silicon integrated devices. Chemical milling can be defined as the controlled etching of selected areas of a given substrate by selective exposure to a suitable etchant.

Lately, a new application field has been added to the classic ones which cannot fail to stimulate interest.

Since the successful flight of Apollo 11 everyone knows that the application of photo-cross-linkable polymers helped the success of this experiment appreciably. A small metal plate, deposited on the surface of our satellite, contains the peace message of 74 nations etched on it with the aid of photoresists. This simple fact convinces us of the definitive entry of these techniques into the interplanetary area [4-6].

During the past few years the chemistry of photo-cross-linkable polymers has been carefully reviewed by several authors [7-10].

Therefore, the task of presenting the recent progress in this field turns out to be in a way greatly simplified.

The first part of this report is devoted to some new synthetic approaches and to new photosensitive polymers. During the past three years the study of the photo-cross-linking mechanism and of the photosensitization has made constant progress. The present state of these problems will be discussed in Section 4.

Recent technological evolution, principally in the field of microelectronics, poses a series of problems which seems not to be solvable by the classic techniques based on known photosensitive polymers. New applications appear to require a constant improvement of the photoresist system proporties. A short survey of some recent developments will be given in Section 6.

2. NEW SYNTHETIC APPROACHES

A great number of photosensitive polymers, among those which gave rise to the most important technical applications, were obtained by substitution of preexisting macromolecules with adequately selected photosensitive derivatives. Within the same scope, the work of Holtschmidt and Oertel opened the way to new sensitive polymers obtained from isocyanates [11-15].

Azide and sulfonylazide groups can also easily be introduced as lateral substituents on polymeric chains. The synthesis of starting materials is shown in Scheme 1.

Carbanionic reaction sequences were used to introduce various photosensitive functions on several new polymers. These techniques were described by Calvayrac and Golé [16]. The basic process is based on the creation of carbanionic centers on a polymeric backbone and on their subsequent reaction with photosensitive halogenated derivatives, as shown in Scheme 2.

Free radical polymerization of vinyl cinnamate does not lend itself to the synthesis of polyvinyl cinnamate, whose excellent photosensitivity is very well known. The presence of two unsaturated groups in the same molecule leads instead to cyclopolymerization and insolubilization reactions.

Kato and Hasegawa recently described the cationic polymerization of p-vinyl-phenyl-cinnamate (I), β-vinyloxyethylcinnamate (II), and β-vinyloxystyrylacrylate (III) [17]. Monomer I was synthetized from p-hydroxystyrene and cinnamoyl chloride. It can be polymerized in methylene chloride solution with boron trifluoride-etherate complex as catalyst. Monomers II and III were synthetized from sodium cinnamate or sodium styrylacrylate and 2-chloroethylvinyl ether. They can be polymerized in dry toluene with the same catalyst. These processes, shown in Scheme 3, allow the preparation of a soluble, highly viscous photosensitive polymer directly from the monomer.

The reaction is free from insolubilization, the cinnamic unsaturation not being influenced by the cationic process. Copolymerization with monomers such as isobutyl-vinyl ether and α-methyl styrene is a very attractive approach since it is obvious that the structure of the photosensitive copolymers can be adapted to the particular requirements such as adhesion to various substrates.

A great number of photosensitive polycondensates have been obtained by

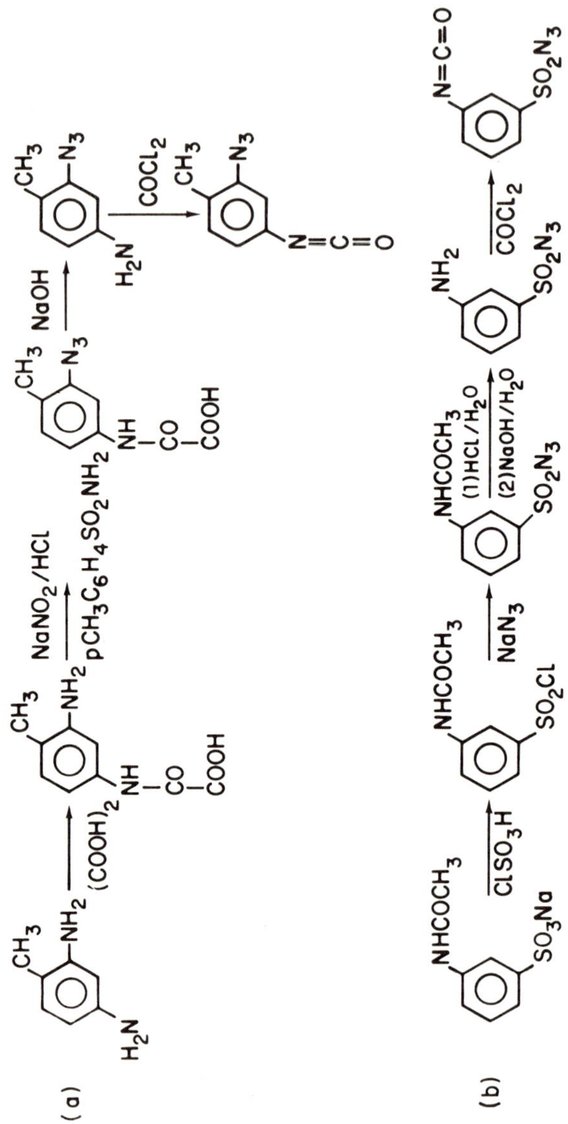

Scheme 1. Synthesis of reactive azido and sulfonyl derivatives.

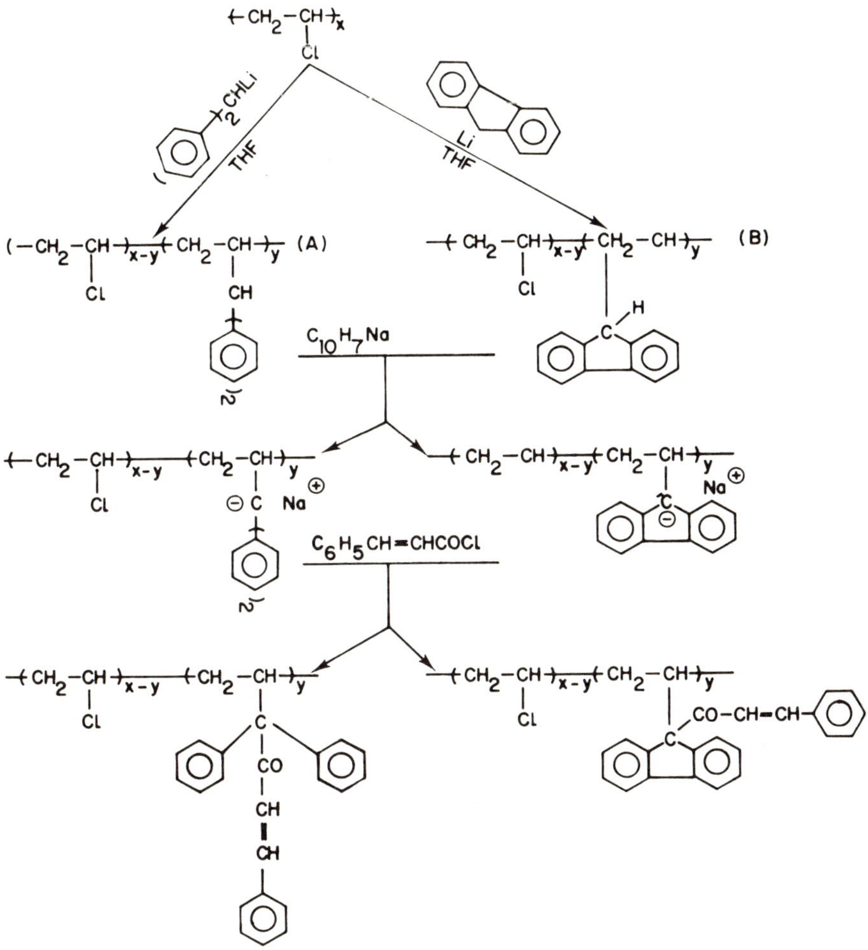

Scheme 2. Synthesis of photosensitive polymers by carbanionic sequences.

interfacial polycondensation techniques. A general survey of the different possibilities is given in Scheme 4.

New light-sensitive polysulfonate copolymers containing the styrylketone group have been prepared by Arcesi et al. [18b]. The copolymers were prepared by the interfacial polycondensation of bisphenol A, divanillalcyclopentanone and m-benzenedisulfonylchloride. The limited solubility of those polymers has been improved by incorporation of several non-light-sensitive bisphenols and aromatic disulfonylchlorides. Replacement of the 1,3-benzenedisulfonylchloride by 1-methyl-2,4-benzenedisulfonylchloride and 4,6-dimethyl-1,3-benzene-

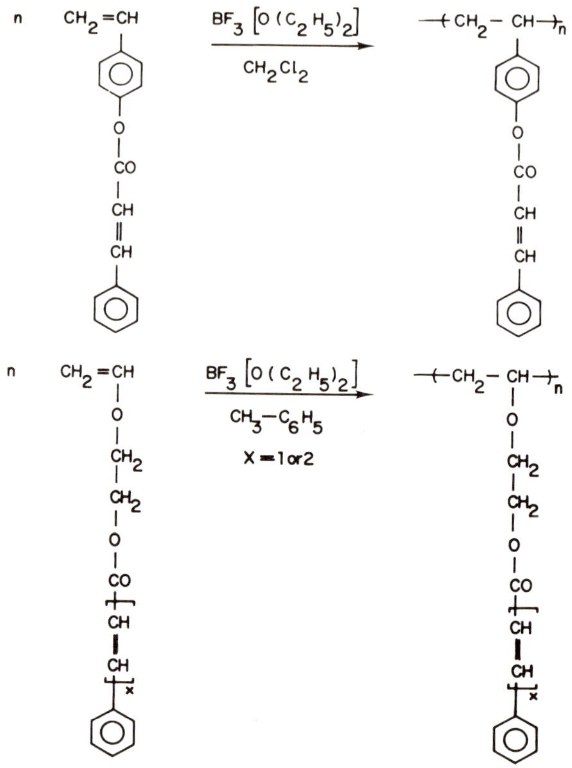

Scheme 3. Cationic polymerization of cinnamic monomers.

disulfonylchloride led to high molecular weight, solubility, and excellent hydrolytic stability to both acid and base etchants.

Especially interesting polyesters obtained from arylazido-substituted acid chlorides were carefully studied. The photolysis of the photosensitive groups and the sensitivity to photochemical cross-linking as a function of degree of substitution and of sensitizer concentration were reported [18a].

3. NEW PHOTOSENSITIVE POLYMERS

During the past few years several new photosensitive polymers were described. However, practically all these new polymers are in the line of preceding structures based on the presence on the polymer backbone of unsaturated dimerizable groups or photolabile functions such as azide groups. Tsuda recently described a very interesting photosensitive polymer based on the introduction of 2-furylacrylic groups on a polyvinyl alcohol chain [19]. The synthesis of this polymer, shown in Scheme 5, takes place at low temperature (−5° to −8°C) in alkaline medium.

The photo-cross-linking mechanism proposed by the author is based on the formation of cyclobutane rings by photocycloaddition of two photosensitive groups from two separate polymer chains. A most interesting aspect of this polymer, when compared to polyvinyl cinnamate, is the remarkable increase of the photosensitivity in the presence of sensitizers. Relative photosensitivity values are given in Table 1.

Table 1. Relative Photosensitivity Values

Polymer	Sensitizer[a]	Photosensitivity
PVC[b]	None	0
PVC	Picramide	400[c]
PVC	Michler's ketone	3,200
PVFA[d]	None	4,400
PVFA	Nitrofluorene	44,550
PVFA	Michler's ketone	89,100

[a]Sensitizer: 10 wt % vs. polymer.
[b]Polyvinyl cinnamate.
[c]Sensitivity with picramide taken as 400.
[d]Polyvinyl-2-furylacrylate.

By proper substitution of the benzene nucleus in the stilbene substituent, Stuber et al. were able to increase in a very important way the photosensitivity of stilbene-containing polymers [20]. Copolymers of maleic anhydride and vinylmethyl ether, and polyhydroxy ethers obtained by polycondensation of bisphenol A and epichlorhydrin, were treated respectively with 4-amino-2′,5′-dimethoxystilbene and 4-isocyanato-2′5′-dimethoxystilbene. Photosensitive polymers carrying substituted stilbene groups attached by amide and urethane bridges to the polymer backbone were also obtained. These syntheses are shown in Scheme 6 where polymer A represents the resulting polymer after esterification of the amide with methanol. As a consequence of the participation of excited singlet states in the photo-cross-linking reaction, those polymers

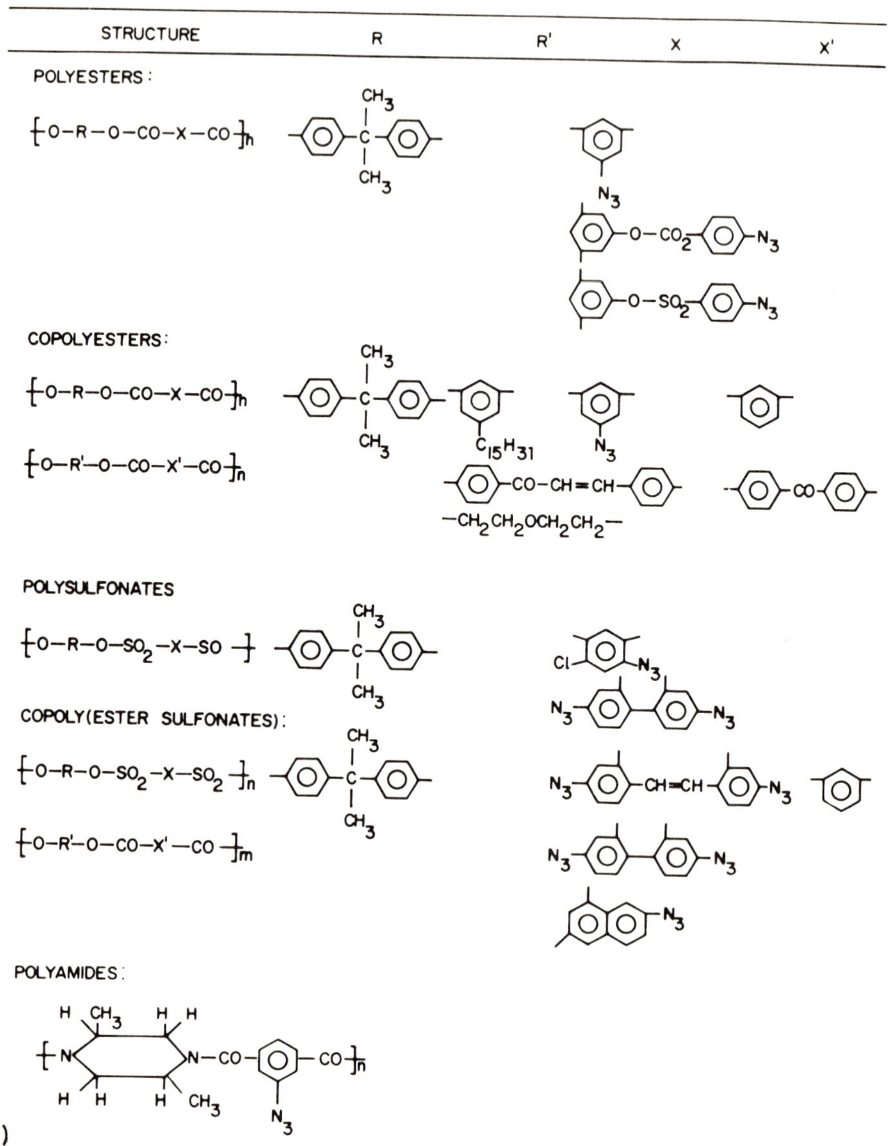

(a)

Scheme 4. (A) Photosensitive aryl azido polycondensates and (B) photosensitive styrylketone polycondensates.

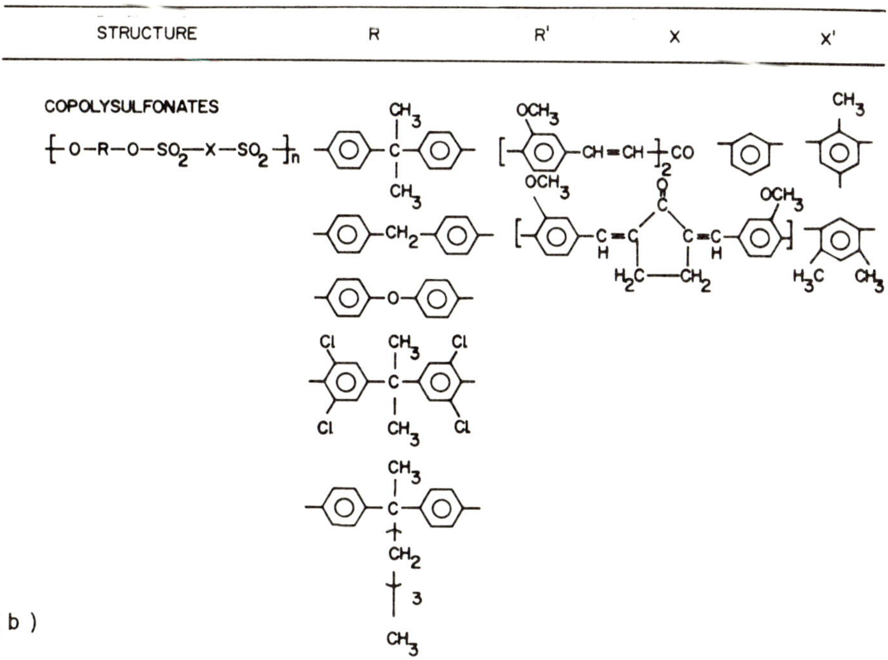

(b)

Scheme 4. (cont.)

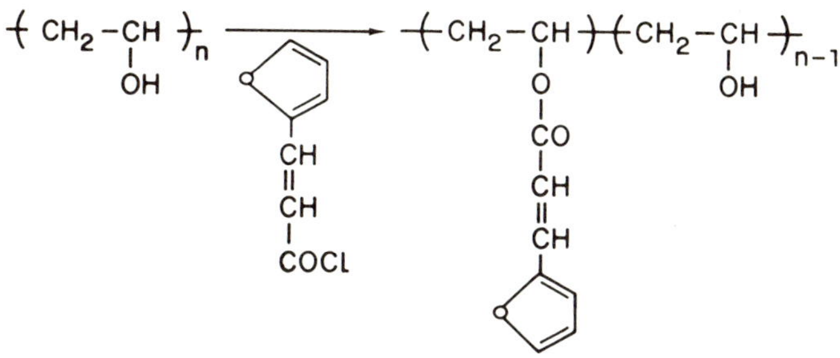

Scheme 5. Synthesis of polyvinyl-2-furylacrylate.

can be photoinsolubilized in the presence of oxygen. The competitive cis-trans-isomerization is at the origin of the better transparency of the films. Because of this behavior, thicker layers can be cross-linked.

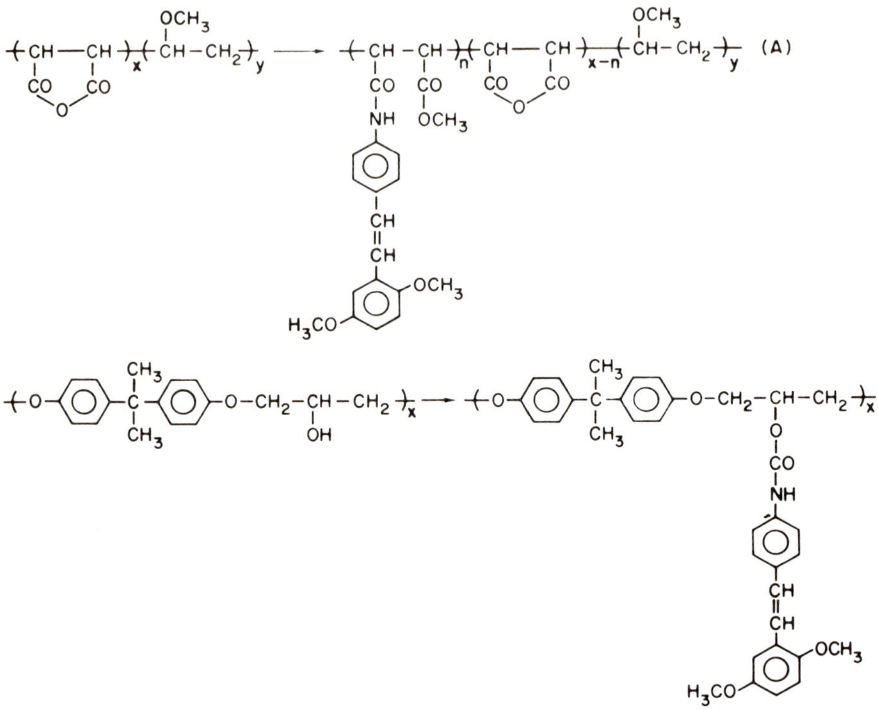

Scheme 6. Synthesis of methoxy-substituted stilbene polymers.

Recently, photosensitive polymers containing conjugated acetylenic bonds were described by Wegner [21].

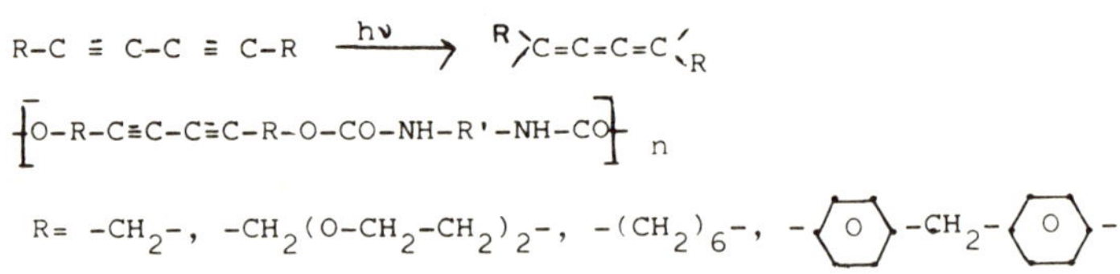

Color formation and insolubilization occur by irradiation. Radical catalysts were shown to accelerate these reactions.

Acetylenic polycondensates were obtained by copper-catalyzed oxidative coupling of bisphenol A-dipropargylether [22]. The monomer was synthetized from bisphenol A and propargylbromide. The process is shown in Scheme 7. Introduction of sensitizing carbonyl groups in the polymer chains markedly

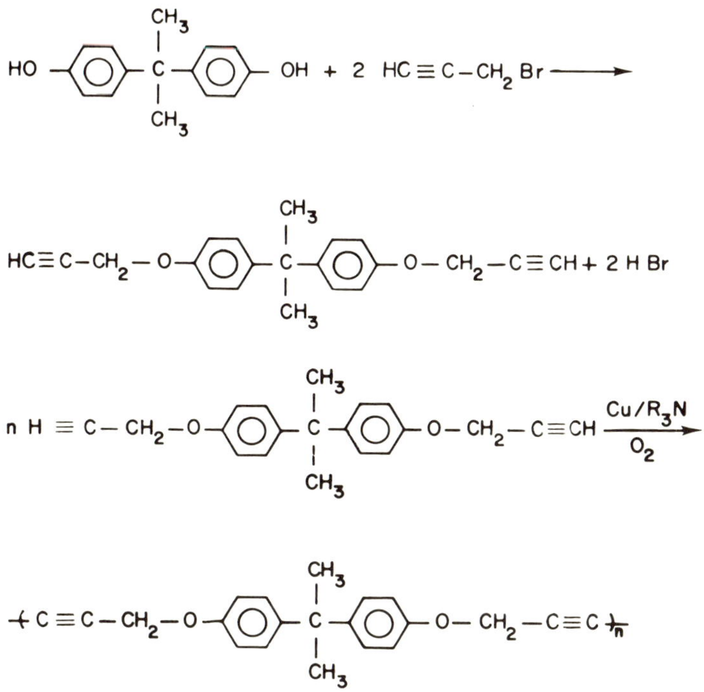

Scheme 7. Synthesis of acetylenic polyethers.

increased the photosensitivity as shown in Scheme 8. The solubility of the polymers was obtained by increasing the flexibility of the chains.

4. PHOTOSENSITIZATION

4.1 Cinnamic Derivatives

The photoinsolubilization mechanism of polyvinyl cinnamate and its photosensitization are without doubt the most thoroughly investigated aspects of the photo-cross-linking of photosensitive polymers. A great amount of work was devoted to these problems.

Two hypotheses were formulated for the photo-cross-linking mechanism. They are based on the photocyclodimerization of cinnamic groups with formation of cyclobutane bridges between polymer chains and on cross-linking

$$\left(C\equiv C-CH_2-O-R-O-CH_2-C\equiv C \right)_n$$

R	CROSS-LINKING TIME (SEC)
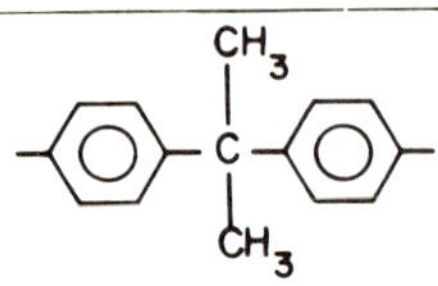	120
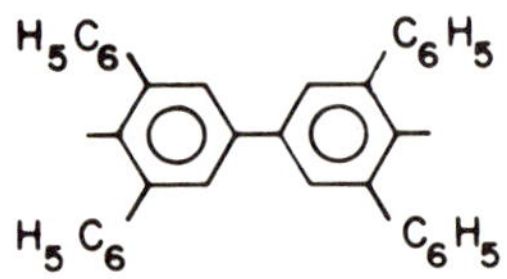	110
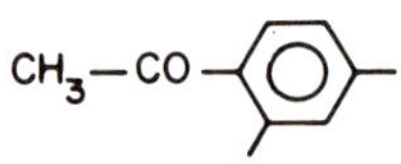	35
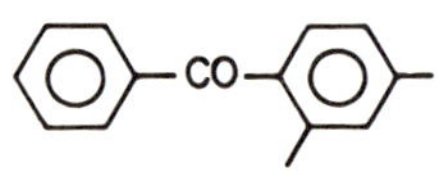	40

Scheme 8. Photosensitivity of acetylenic polyethers.

resulting from free radicals formed during the irradiation. Highly valuable arguments were put forward by several authors in favor of both hypotheses.

The photochemical formation of cyclobutane-type dimers from cinnamoyl groups was recently described as a one-step four-membered ring formation by Tsuda [23]. This reaction relates to the cyclobutane formation from ethylene described from the point of view of conservation of orbital symmetry by Woodward and Hoffmann [24].

The exact mechanism of the incipient photo-cross-linking was not unambiguously demonstrated, however. Simultaneous participation of both mechanisms seems to be highly probably. The participation of the excited triplet state of the cinnamic group in the photo-cross-linking reaction was nevertheless clearly demonstrated. This particular excited state can be produced by direct irradiation and intersystem crossing from the excited singlet state or by triplet energy transfer from a properly selected sensitizer.

Some recent reports corroborate this hypothesis irrefutably. The work of Moreau was primarily devoted to the photosensitization of polyvinyl cinnamate by 2-benzoyl-methylene-1-methyl-naphthothiazoline [26]. This author showed clearly that the fluorescence of this sensitizer was not at all affected by the presence of polyvinyl cinnamate, whereas the phosphorescence quantum yield was decreased from 0.85 to 0.15. In the same work several important criteria for the selection of active sensitizers were described: (a) the presence of a triplet state at an energy level close to the same state of the cinnamic group located at 50-55 kcal/mole, (b) a phosphorescence to fluorescence quantum yield ratio higher than unity, and (c) a triplet state mean lifetime greater than 0.01 sec.

Keeping these rules in mind, spectroscopic data from flash photolysis studies and spectrofluorimetric measurements can be used to establish a photosensitivity scale for a great number of sensitizers as shown in Table 2. The triplet state energy value of 4-nitro-1-naphthylamine and 4-nitroaniline were taken from the work of Foster et al. [27].

Nakamura and Kikuchi also reported on the photosensitization of polyvinyl cinnamate [28]. They were able to show the phosphorescence of cinnamic groups at about 20,000 cm^{-1} and the photosensitizing activity of derivatives characterized by a triplet state located at or beyond this energy level. They proposed a Jablonski energy diagram describing the photosensitization of polyvinyl cinnamate as shown on Scheme 9 for 2-nitrofluorene. Moreover, these

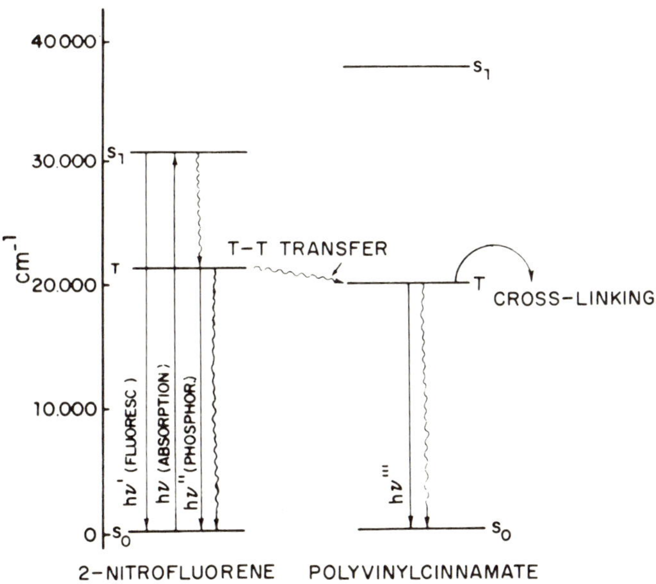

Scheme 9. Energy transfer diagram for polyvinyl cinnamate photosensitization.

authors determined the critical distance for sensitization whose existence was clearly related to the energy transfer mechanism by exchange between triplet states.

Table 2. Photosensitization and Spectroscopic Characteristics

Sensitizer	Relative speed[a]	E_T, kcal/mole	τ, sec[b]	ϕ_P/ϕ_F[c]	ISCE[d]
Anthrone	31	72	0.01	1000	–
Benzanthrone	7	72	0.001	1000	–
Benzophenone	20	70	0.001	1000	1
Phenanthrene	14	62	3.3	1.1	0.76
Naphthalene	3	61	2.3	0.11	0.39
Michler's ketone	640	61	6	1000	1
2-Nitrofluorene	113	59	–	–	–
4-Nitrobiphenyl	200	58	0.5	1000	–
5-Nitroacenaphthene	84	56.6	–	–	–
Chrysene	18	56	2.5	0.51	0.67
4-Nitroaniline	100	55	0.1	1000	–
2-Nitroaniline	0	–	–	0	–
Naphthothiazoline	750	54	0.5	100	–
N-Acetyl-4-nitro-1-naphthylamine	1100	52.5	–	–	–
Polyvinyl cinnamate	1	50-55	–	–	–
4-Nitro-1-naphthylamine	0	49.4	–	–	–
1,2-Benzanthracene	0	47.8	–	–	–
Anthracene	0	42	0.01	1.5	–

[a]Relative photosensitivity with respect to unsensitized polyvinyl cinnamate.
[b]Triplet lifetime.
[c]Phosphorescence to fluorescence quantum yield ratio.
[d]Intersystem crossing efficiency.

It is advisable to mention here the conclusive experiments of Tsuda [29]. Substitution by heavy atoms of sensitizer molecules allowed a remarkable increase of the phosphorescence quantum yield owing to spin-orbit interaction. The consequence of such a drastic change was a higher photosensitivity aptitude of sensitizers with correctly located triplet states. Furthermore, this author was able to transform a molecule devoid of sensitizing properties into a powerful sensitizer by correct substitution. This modification of sensitizing aptitude is based on the shift of the triplet energy to an efficient level when going, for example, from 4-nitro-1-naphthylamine to its N-acylated derivatives. The relative photosensitizing speeds obtained by Tsuda are summarized in Table 3.

Table 3. Relative Photosensitizing Speeds

Sensitizer	Relative speed[a]
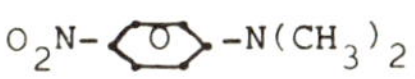	137
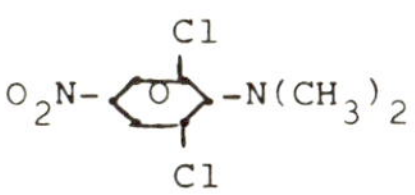	561
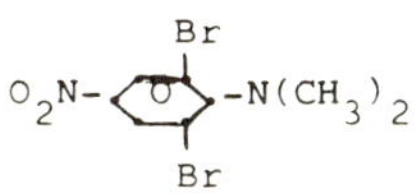	797
	0
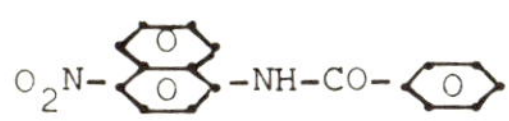	769
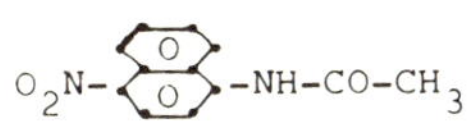	1100

[a]Relative speed for unsensitized polyvinyl cinnamate taken as unity.

4.2 Azido Derivatives

Excellent photo-cross-linkable polymer compositions were based on the photolytic ability of azido derivatives. During this photodissociation reaction very reactive fragments are formed. Nitrenes, characterized by a monovalent nitrogen atom, appear after the elimination of a nitrogen molecule.

These reactive groups are rapidly deactivated along several reaction paths very well known to the organic photochemist. Some reactions can give rise to insolubilization of the irradiated polymers. The utilization of azido derivatives in photo-cross-linkable polymers appeared mainly in formulations obtained by the synthesis of azido-substituted polymers and by the addition of aromatic diazides to properly selected polymers. These techniques afforded the development of very sensitive resist formulations.

The study of the azide photolysis and that of its photosensitization has made great progress during the past three years. The work of Reiser and coworkers allowed a better understanding of the aromatic azide photolysis [25, 30-32].

The most intense bands of the aromatic azides spectra were assigned to π-π^* transitions. The general features of the hydrocarbon spectra are maintained although the presence of the azide group reduces the symmetry of the molecule. Nearly all azide spectra show a low-lying shoulder on the long wavelength side of the ^1L_b band. These bands were assigned to n-π^* transitions of a lone pair electron from nitrogen into the aromatic π system. The spectra of the nitrenes are quite different from those of the parent azides. The nitrenes have one unpaired aromatic electron and their spectra are similar to those of aromatic radicals. Only π-π^* transitions associated with the aromatic system have been observed so far. The transients in the flash photolysis of several aromatic azides were identified as triplet nitrenes.

A careful study of the photolysis quantum yields showed a dependence of the aromatic azides photolysis on the structure of these derivatives. The photolysis of an aromatic diazide requires the absorption of two photons. Several results are summarized in Table 4. Among the three possible dissociation modes of the azide group only the vibrational excitation of the critical bond was considered to be valid. As a consequence of this the photolysis quantum yield should be dependent on the irradiation wavelength. Aromatic azides do not show fluorescence emission when irradiated by ultraviolet light. Very efficient internal conversion occurs and dissociation of the molecule follows practically every transition from the excited (n-π^*) state to a high vibration level of the ground state.

When excited to a (π-π^*) level the molecule crosses over to the (n-π^*) state before dissociation. Very efficient energy transfer from light-absorbing groups to RN—N_2 bonds were observed when a carbonyl function was present in the molecule. This behavior explains the particularly efficient photo-cross-linking activity of these carbonyl-substituted derivatives.

The absorption of less energetic photons places less energy at the disposal of the vibrational dissociation. Derivatives such as

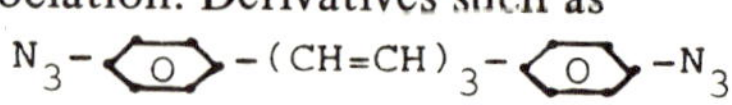

seem indeed to undergo a less efficient photolysis. These hypotheses are based on excited singlet states. Many questions remain unanswered, however. They are mainly concentrated on the existence of excited triplet states and dissociative excited states and on the exact nature of the nitrene intermediates.

Table 4. Aromatic Azide Photolysis – Quantum Yields

Azide	λ, nm[a]	ϕ_1[b]	ϕ_2[c]
	254	0.53 ± 0.1	–
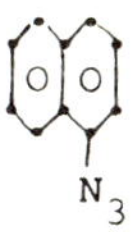	303	1 ± 0.02	–
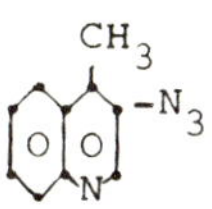	303	0.84 ± 0.05	–
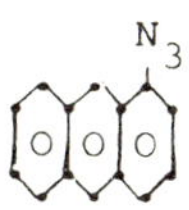	365	1 ± 0.02	–
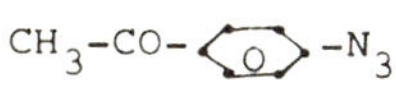	280	1 ± 0.05	–
	365	1 ± 0.01	–
	254	0.45 ± 0.1	1 ± 0.1
	304	0.45 ± 0.1	1 ± 0.1
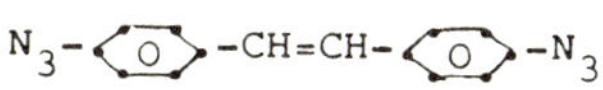	365	0.37 ± 0.05	1 ± 0.05
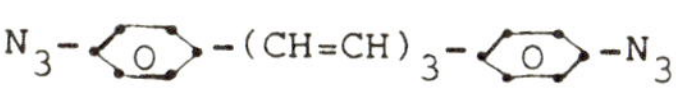	405	0.2 ± 0.1	0.6 ± 0.1
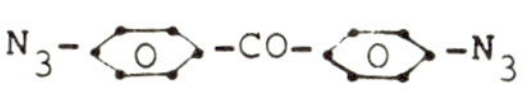	303	1 ± 0.1	–
	365	1 ± 0.1	–

[a]Irradiation wavelengths.
[b]Quantum yield for the first azide group.
[c]Quantum yield for the second azide group.

The photosensitization of the azide photolysis is also a very actual problem. The technological importance of extending the photosensitivity of photo-cross-linkable systems as far as possible into the visible part of the spectrum requires the use of photosensitizers and an exact knowledge of the photosensitizing mechanism. Up to now the choice of the sensitizers was implicitly influenced by the hypothesis of a triplet-triplet energy transfer mechanism. Recently, however, the use of new polyaromatic and heterocyclic sensitizers was reported [33]. This allowed a reconsideration of the triplet hypothesis. Among the polyaromatic derivatives, anthracene and its substitution products were shown to be particularly interesting. The excited triplet state of anthracene is situated at 42 kcal/mole whereas the probable location of the aromatic azide excited triplet state must be at about 75 kcal/mole, according to Lewis and Saunders [34].

A triplet-triplet energy transfer seems to be highly improbable because of the strong endothermic character of this particular transfer. The hypothesis of an energy transfer between singlet states must therefore be seriously considered. As a consequence of this transfer, a quenching of the sensitizer fluorescence should be observed. The fluorescence of anthracene was as a matter of fact quenched by the addition of an aromatic azide or an aromatic azido-substituted polymer. This quenching can be shown to obey a Stern-Volmer equation as represented in Fig. 1 [35].

Assuming the mechanism (I-IV), where the rate of production of the fluorescent singlet state of anthracene (A*) is constant and equal to the product of the light intensity, I, and the efficiency, f, of the light absorption, and where k_2 represents the sum of the rate constants for decay of singlet excited anthracene by all other processes except fluorescence and quenching by azides (B), the ratio of the fluorescence quantum yields in the presence (ϕ_A) and in the absence of quencher (ϕ_0) was easy to obtain experimentally.

$$A \xrightarrow{h\nu} A^*; \quad fI \qquad \text{(I)}$$

$$A^* \longrightarrow A + h\nu' \quad ; k_1(A^*) \qquad \text{(II)}$$

$$A^* \longrightarrow A \; ; k_2(A^*) \qquad \text{(III)}$$

$$A^* + B \longrightarrow A + B^* \quad ; k_3(A^*)(B) \qquad \text{(IV)}$$

The decay rate is given by the following expression:

$$\frac{d(A^*)}{dt} = f\,I - k_1(A^*) - k_2(A^*) - k_3(A^*)(B)$$

For steady state conditions this equation can be written as follows:

$$fI = [k_1 + k_2 + k_3(B)]\ (A^*)$$

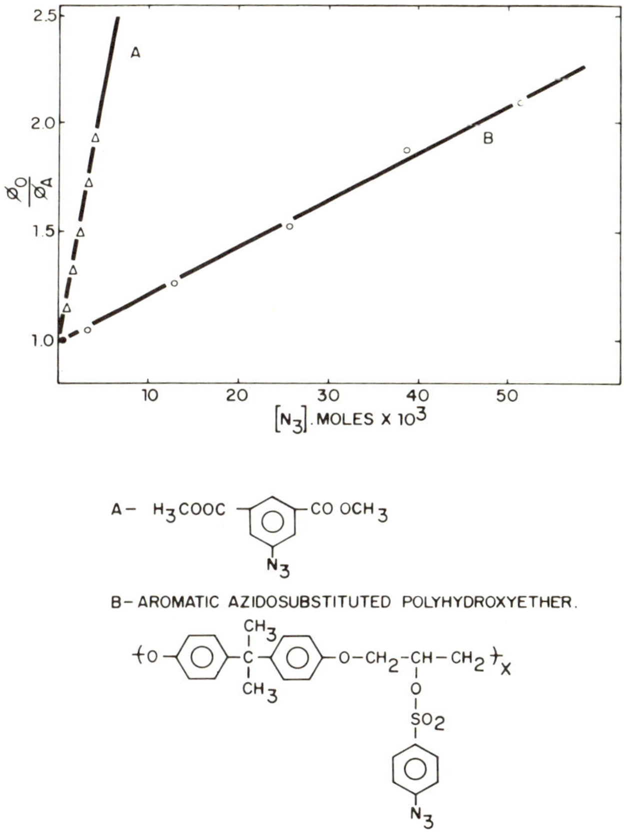

FIG. 1. Quenching of anthracene fluorescence (403 nm) by aromatic azido derivatives.

Taking into acount the quantum yield relationships the Stern-Volmer equation can be deduced.

$$\phi_A = k_1(A^*)/fI = k_1/k_1 + k_2 + k_3(B)$$
$$\phi_0 = k_1/k_1 + k_2$$
$$\phi_0/\phi_A = 1 + k_3/k_1 + k_2 \cdot (B) = 1 + k_3\,\tau(B)$$

In this expression the lifetime of the excited anthracene singlet state (τ) is given by $1/k_1 + k_2$. Replacing (B) by the quencher concentration $(B)_{1/2}$ required to quench 50% of the excited molecules and assuming a constant value of 10^{-8}

sec for the singlet anthracene lifetime, the quenching rate constants k_3 could be calculated.

The following values of k_3 were found: $k_3 = 2.5 \times 10^{10}$ liter mole^{-1} sec^{-1} for quenching by 3-azido-dimethylisophthalate and $k_3' = 2.1 \times 10^9$ liter mole^{-1} sec^{-1} for quenching by the azidopolyhydroxy ether. Inasmuch as the relative viscosity ratio of the polymer solution to the 3-azido-dimethylisophthalate solution, both in methylene chloride, has a value of 2.7, this tenfold decrease of the quenching rate constant by adding a polymeric azide quencher could not be explained on the single basis of a viscosity effect on a pure diffusion-controlled process. The rather high value of k_3 can however be correlated with a long-range Förster-type energy transfer. The rate decrease observed for k_3' could be attributed to a shielding of the quenching groups by the polymer coil giving rise to a higher critical quencher concentration $(C_0)_{1/2}$ for the long-range transfer.

Sensitization by excited singlet states responsible for the fluorescence seems to be corroborated by experiments with heterocyclic oxazole derivatives. Some sensitizers such as 2,5-diphenyl-oxazole are known for their high fluorescence quantum yields. They are moreover excellent sensitizers as can be seen in Table 5 [35].

Substitution of two azido groups on an oxazole-type heterocycle gave rise to highly valuable photo-cross-linking sensitizers. Photosensitization of aromatic azide photolysis by singlet energy transfer seems consequently to have to be taken into consideration.

The hypothesis of a singlet-singlet energy transfer for azides was confirmed recently by Moriarty and Reardon, who showed chrysene (E_T = 56 kcal/mole) to act as a photosensitizer for the photolysis of alkyl azides [36]. Fluorescence quenching followed by azide photolysis occurred when hexyl azide was in the presence of singlet-excited chrysene. Swenton et al. reported on the utilization of aromatic hydrocarbons as sensitizers for the photodecomposition of 2-azido-biphenyls as a result of singlet energy transfer [37]. Pyrene (E_T = 48.7 kcal/mole) has been utilized in a unique sense. It functions to populate the azide-excited singlet state by singlet-singlet energy transfer and also serves as a quencher for the excited triplet azide.

Tsunoda et al. succeeded in the esterification of more than 90% of hydroxy groups of polyvinyl alcohol by reacting with p-azido-benzoylchloride. They were able to extent the sensitivity of poly(vinyl-p-azidobenzoate) (PVAB) to 500 nm [38]. Several sensitizers were used and their relative sensitivity values compared with poly(vinyl cinnamate) (PVC) are shown in Table 6.

Particularly interesting systems were obtained by using internally sensitized diazides as photosensitive cross-linking agents for cyclized rubbers [39,40]. 2,6-

Table 5. Relative Photosensitivity Speeds for Heterocyclic Derivatives[a]

Sensitizer[b]	Relative speed[c]
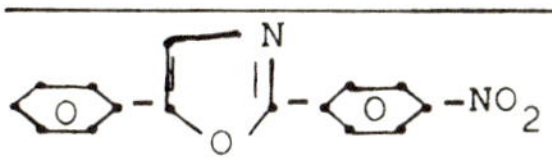	1
	1.4
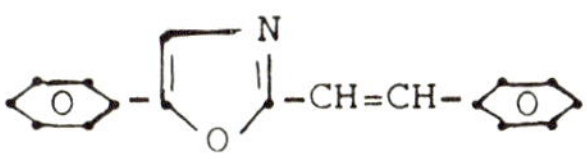	2
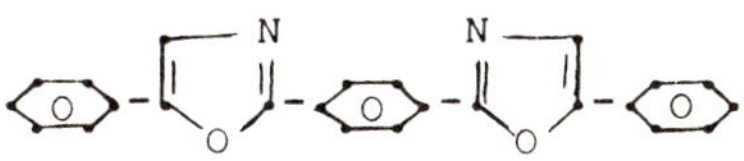	0.35
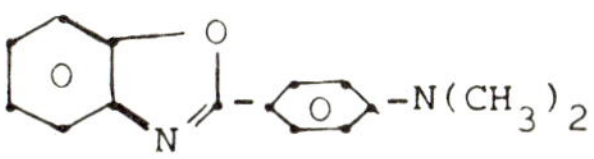	2

[a]Photosensitive polymer:

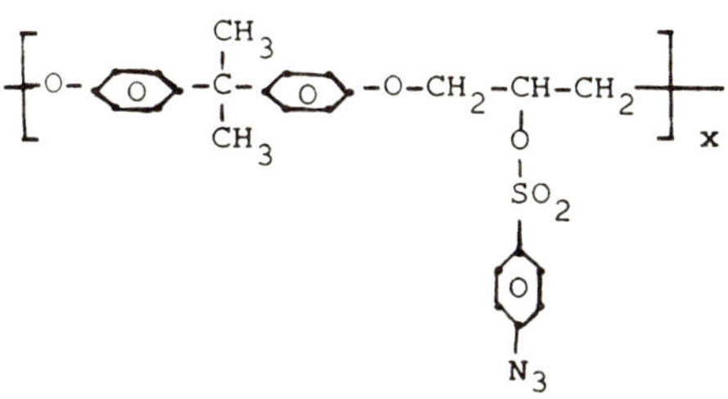

[b]Sensitizer concentration: 10 wt % vs. polymer.
[c]Photosensitivity speed with Michler's ketone taken as unity.

Bis-(4′-azidobenzal)-methylcyclohexanone and 2,6-bis-(4′-azidobenzal)-cyclohexanone appeared to be powerful photo-cross-linking agents. The sensitivity value was shown to undergo a tenfold increase when compared to the unsensitized 4,4′-bis-azidodiphenylmethane.

The same internal carbonyl sensitization was described by Mistr et al. [41]. These authors were able to show a marked increase of the light sensitivity of the 4,4′-diazidodibenzalacetone as compared to the 1-4-bis-(4′-azidobenzoyl-oxy)-benzene.

Table 6. Sensitivity and Spectral Response of Poly(vinyl-p-azidobenzoate)

Poly-mer	Sensitizer	Relative sensi-tivity	Sensitized speed value	Wavelength range, nm	
				Unsensitized	Sensitized
PVC	–	1	–	260-340	–
PVC	1.2-Benzanthraquinone	–	231	–	260-470
PVAB	–	2000	–	220-390	–
PVAB	1.2-Benzanthraquinone	–	338,381	–	390-480
PVAB	5-Nitroacenaphthene	–	120,045	–	390-490
PVAB	N-Phenylthioacridone	–	170,000	–	390-520

5. MONORADICAL PROCESSES

A great number of photo-cross-linking processes involving cross-linkable polymers carrying internal or lateral unsaturation and polymers carrying photo-labile fragments giving macromolecular-initiating free radicals are described.

5.1. Polymers with Lateral (Side Chain) Unsaturation

Isolated macromolecules carrying unsaturated groups either in a lateral position or as an integral part of the chains can be very easily cross-linked in the presence of photolytically formed free radicals. Although these polymers possess a certain intrinsic sensitivity, the addition of sensitizers is generally required. Different possibilities are described in the literature [9]. Recently, interesting unsaturated polymers were obtained by substitution of hydroxyl groups in polyhydroxy ethers with acid chlorides or methylolacrylamide [42]. The general structure of these polymers is given by:

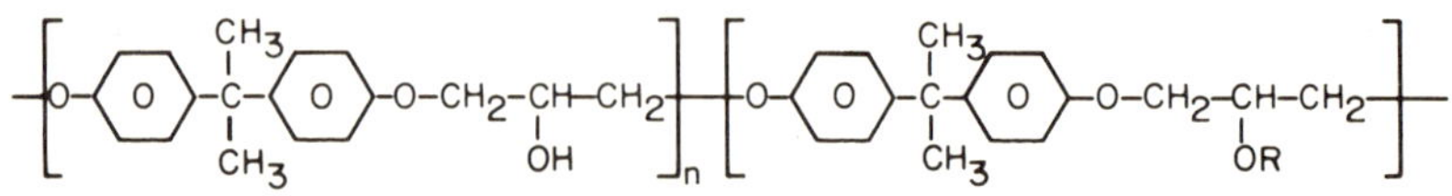

where

$$R = \text{—CO—CH=CH}_2,\ \text{—CO—}\underset{\displaystyle CH_3}{\underset{|}{\text{C}}}\text{=CH}_2,\ \text{—CH}_2\text{—NH—CO—CH=CH}_2,$$

$$\text{—CO—CH=CH—COOH}$$

The same procedure was applied with several hydroxy-substituted polymers such as partially hydrolyzed copolymers of vinyl acetate.

Different free radical catalysts were used for the photo-cross-linking of these polymers. Among them benzoin, benzoin-methyl ether, benzophenone, and benzil were found to be most efficient. An important synergistic effect was observed when Michler's ketone was added to one of these sensitizers. The sensitivity of the acryloyl-substituted polyhydroxy ether in the presence of a mixture of benzoin and Michler's ketone was shown to extend up to 420 nm as can be seen in Fig. 2 where the logarithm of the light sensitivity is plotted against the irradiation wavelengths. More recently, Gilano and Lipson reported

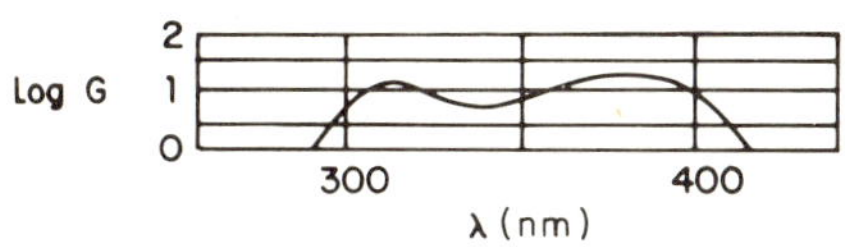

FIG. 2. Acryloyl-substituted polyhydroxy ether photosensitivity: G = I/E and E = It (exposure) (einsteins/$cm^2 \times 10^8$).

the synthesis and photo-cross-linking of homopolymers and copolymers of polyfunctional allyl esters containing two or more unsaturated groups (43).

Polymerization of diallylisophthalate with peroxide catalysts was investigated. This polymerization is an extremely complex process. At approximately 20% conversion the soluble β-prepolymer can be isolated and further used for photo-cross-linking purposes. This prepolymer is a relatively stable material exhibiting excellent properties of dimensional and thermal stability, chemical resistance, adhesion, and electrical resistence.

The β-prepolymer can be cross-linked with aryl diazides and with aryl ketones.

Synergetic effects can be obtained by combination of sensitizers at optimized ratios. Sensitization with mixtures of benzil or benzophenone with Michler's ketone allows for efficient cross-linking in the 3200-4000-nm range. This synergism is quite similar to the particular behavior reported earlier for acryloyl-substituted polyhydroxy ethers.

5.2. Polymers with Photolabile Groups

Polymers carrying dithiocarbamate or xanthate lateral substituents were described [9]. They were able to initiate photografting reactions and to be insolubilized in the presence of bifunctional monomers and sensitizers such as benzoin methyl ether. These polymers act not merely as photoinitiators but also as powerful chain transfer agents as can be concluded from the considerable cross-linking observed during photografting.

Many O-acylated oximino ketones have recently been synthetized from the corresponding α-β-diketones. These derivatives can easily be photolyzed, and the free radical fragments produced during the photolysis initiate the polymerization of acryl derivatives [44].

2;3-Butanedione-O-methacroyloxime was copolymerized with methyl methacrylate in acetone solution at various molar ratios using azobisisobutyronitrile as initiator.

The structure of the copolymers can be represented as follows (where x/y = 13, 5.5 and 2.5):

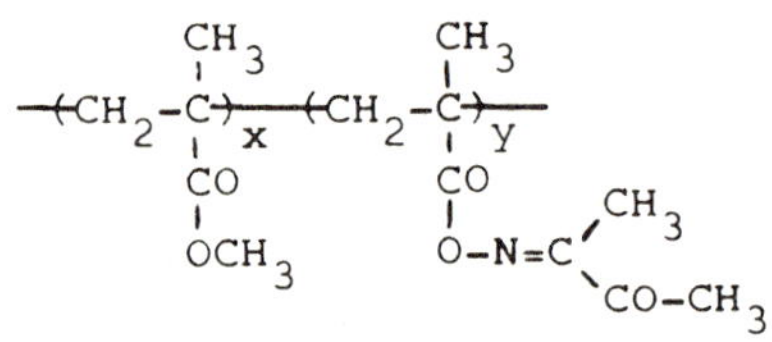

These copolymers were further used as macro-initiators for the photochemical grafting of styrene and acrylamide, respectively, in benzene and dioxane solution. The use of bi- and multifunctional monomers gives rise to photo-cross-linking reactions of the ketoxime copolymers.

Studies on the polymerization mechanism of several monomers containing free radical sensitive groups have indicated that chain transfer reactions may result in cross-linking. Benzyl ethers are particularly sensitive to free radical attack.

Symcox and Cotman described the synthesis and the photo-cross-linking of p-vinylbenzylmethyl ether [45]. This monomer was obtained by refluxing p-(β-chloroethyl)benzylchloride with excess methanolic potassium hydroxide for 3 hr and was polymerized at 50°C with classic free radical initiators. Extensive cross-linking was observed in photoinitiated polymerization with biacetyl as photoinitiator.

Benzoin is very well known as a powerful photoinitiating molecule. To obtain a new type of photosensitive polymer containing a benzoin radical, benzoin acrylate was first synthetized by the reaction of acryloyl chloride with benzoin in dimethylaniline [46].

The monomer was polymerized in dioxane with azobisisobutyronitrile as an initiator. Copolymerization with styrene and with methyl methacrylate was also carried out. Photopolymerization of styrene and methyl methacrylate in the presence of poly(benzoinacrylate) proceeded fast and graft and cross-linked polymers were obtained in good yield. Rapid photo-cross-linking of poly(benzoinacrylate) and benzoinacrylate-styrene copolymer films occurred when irradiated with light of wavelength below 320 nm corresponding to a n-π^* transition of the carbonyl group.

Among the sensitizers, p-benzoquinone was the most effective in causing rapid insolubilization. Kurusu et al. described a new type of photosensitive polymer based on the presence of furyl groups on the polymer chain [47]. Furoin acrylate was obtained by the reaction of acryloyl chloride with furoin and was polymerized or copolymerized with styrene in the presence of azobisisobutyronitrile. Several new synthetic approaches were carried out and are represented schematically in Scheme 10.

The photosensitivity of poly(furoin acrylate) was shown to be higher than the photosensitivity of poly(benzoinacrylate) when irradiated with a high pressure mercury lamp, probably because the former is more effective to longer wavelengths.

Photochemical cross-linking systems based on zero-valent metal complexes such as manganese decacarbonyl and polyvinyltrichloroacetate with added monomers were described by Bamford et al. [48].

5.3. Photoredox systems

Vinyl monomers can be polymerized by irradiation with visible light in the presence of dyes such as eosin, safranine, or acridine yellow, and suitable reducing agents such as ascorbic acid and oxygen [49]. In some cases the incorporation of dyestuff fragments in the polymer chain was observed.

Smets et al. were able to synthesize polyvinylamine with lateral eosine fragments bound through amide links [50]. These polymers were used for the photografting of styrene and acrylamide. The eosine fragments are photoreduced to leuco-eosine by the reducing agents. In the presence of oxygen the leuco dye can be reoxidized. Semiquinone radicals, appearing as partially reduced intermediates, initiate the grafting of the vinyl monomers. In the presence of bi- or multifunctional monomers photo-cross-linking can be expected. The same reaction pattern was proposed by Manecke and Kossmehl

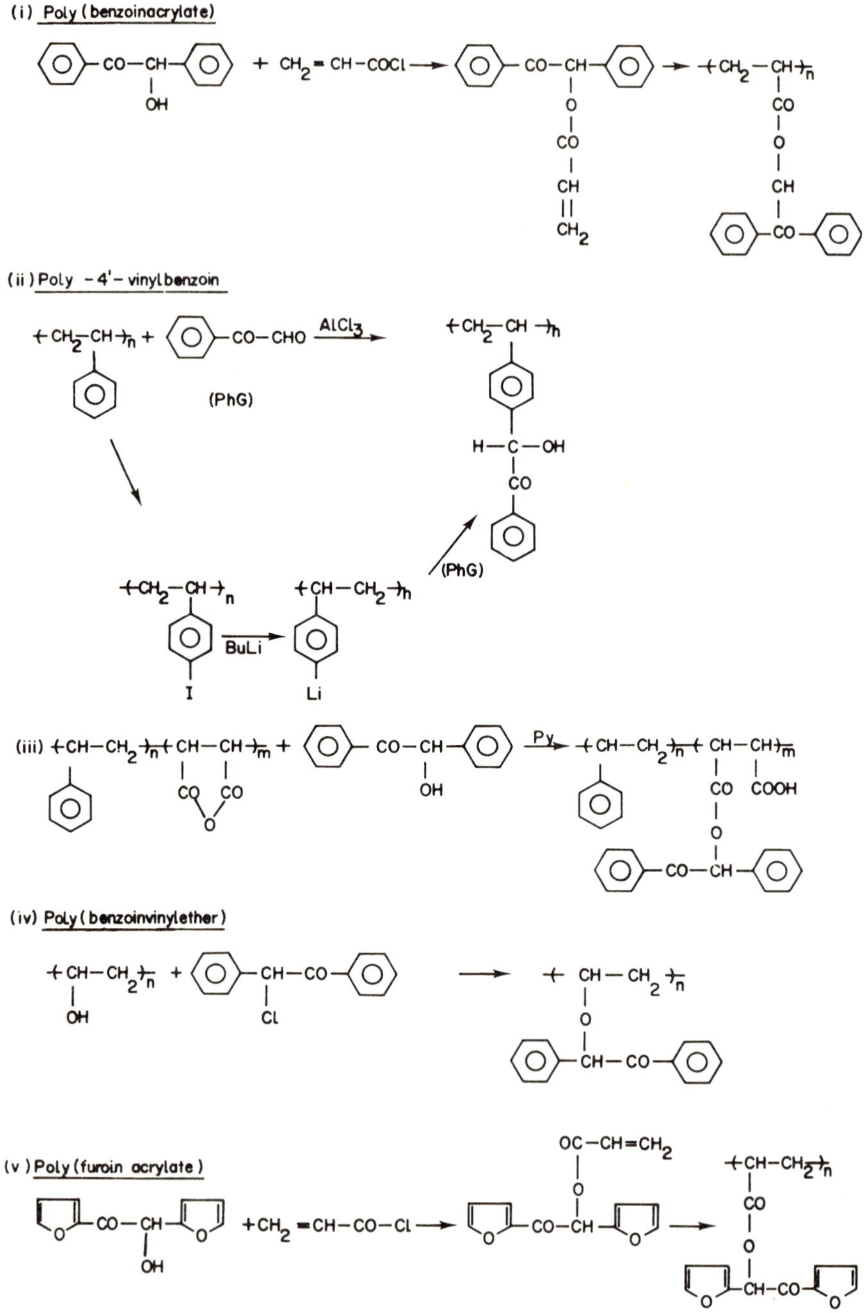

Scheme 10. Synthesis of benzoin and furoin polymers.

for the photoinitiated grafting of styrene with poly(6-vinylisatin), poly(6-vinylindirubin) and poly [6-vinylindol-(3)-] - [thionaphthen(2)] -indigo in the presence of ascorbic acid or thiourea as reducing agents [51-52]. The synthesis of these redox polymers is shown in Scheme 11.

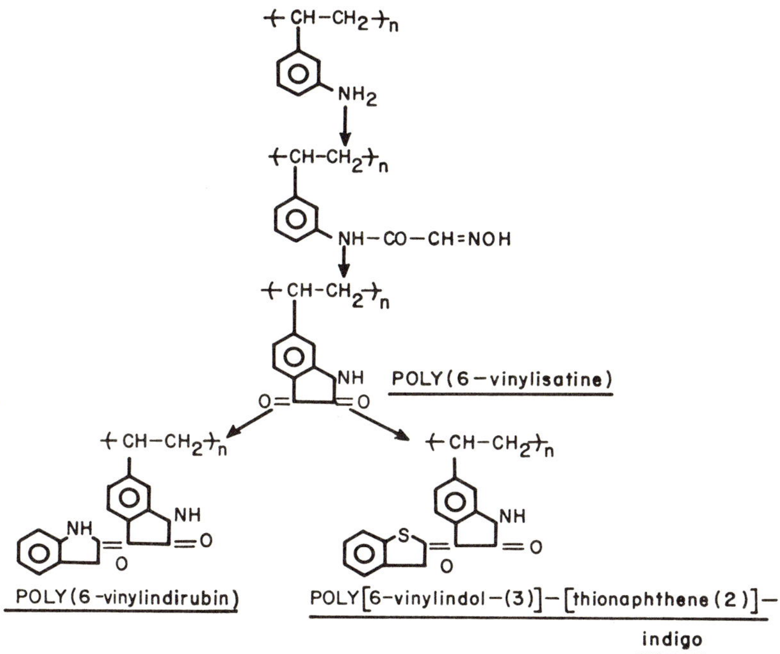

Scheme 11. Synthesis of redox polymers.

6. FUTURE TRENDS

Utilizers of photoresist systems are currently struggling with a certain number of problems. New substrates introduce new adhesion problems whose solution is still limited by our understanding of the adhesion mechanism. Smaller windows in silicon wafers presuppose higher resolution. Those problems and anticipated solutions are reviewed by Lussow [53].

The present trends on microelectronics is toward large-scale integration (LSI) where a small size and a large number of gates are required on a single wafer. In the manufacture of integrated circuits the best resolution obtainable using optical techniques is limited by the diffraction of light and the characteristics of the photographic mask used. The use of electron beams in photoresist operations overcomes the limitations imposed by the use of optical light sources. Power densities are several orders of magnitude higher and resolution can be improved by focusing an electron beam to 0.1 μ with a depth of focus of 25 μ. New problems arise however from the need for a high vacuum system. The response of resists to electrons can be increased by an order of magnitude without any loss of resolving power if certain additives are incorporated in the resist solution [54-56]. A photoresist film can be insolubilized as well with electron beams as with ion beams. The irradiation characteristics for 30-50 keV argon ions are very similar to the characteristics for electron beams [57].

Polymer films can also grow on surfaces hit by ion beams when polymerizable organic molecules are present in the gas phase. The molecules must primarily be adsorbed on the irradiated surface [58].

Formation of resist patterns were also observed when halogenated monomers such as hexachlorobutadiene or tetrafluoroethylene were irradiated with UV in the gas phase through a suitable mask. Photopolymerization occurs at the irradiated surface and produces a very thin pinhole free resist [59].

Selective etching by photochemical in situ formation of etchant was described [60,61]. Polymeric films on a butyrate base, containing N-chloro-, N-bromosuccinimide, N,N′-dibromodimethylhydantoin or iodoform as photolytical halogen formers are used for gold etching. In the case of silicon oxide layers use was made of a polystyrene film containing trityldifluoramine. Hydrogen fluoride was formed by reaction of photolytically formed fluorine with polystyrene [62].

Etching of selected areas of a thin film of p-xylene polymer on silicon by a gaseous electrical glow discharge in nitrogen, argon, or oxygen was also reported [63]. In view of resist formation by projection exposure more sensitive photoresists are needed with spectral response to higher wavelengths. The photosensitivity of photoresist systems toward argon-laser irradiation was studied in view of their utilization as holographic recording materials [64]. Resist formation by scanning techniques without masks, laser exposure, and elimination of the target are also mentioned. Ionized dopant atoms accelerated to high velocity and directed toward the material to be doped would finally eliminate the need for photoresists. The development of the first automatic, nonliquid photoresist for the etching and the plating of printed circuits has also been reported [65].

REFERENCES

[1] G. Oster, *Encyclopedia of Polymer Science and Technology,* Wiley (Interscience), Vol. 10, 1969 p. 145.

[2] H. Gersheim, *History of Photography,* Oxford Univ. Press, New York, Chap. 5, 1966.

[3] A. Murray, US. Pat. 1,965,710 (July 1934).

[4] D. Hemenway, *Photomethods Ind.,* **12**, 45-72 (1969).

[5] I. D. Burr, *Microphot. SPSE-Symposium,* **3**, 139-144 (1968).

[6] *Brit. J. Phot.,* **116**, 981 (1969)

[7] J. L. R. Williams, *1st Retec Conf., Tech. Papers,* 123 (1967).

[8] J. L. R. Williams, *Fortschr. Chem. Forsch.,* **13** (2), 227 (1969).

[9] G. A. Delzenne, *European Polymer J. Suppl.* 55-91 (1969).

[10] J. F. Rabek, *Photochem. Photobiol.,* **7**, 5-57 (1968).

[11] H. Holtschmidt and G. Oertel, *Angew. Makrom. Chem.,* **9**, 1 (1969).

[12] Agfa-Gevaert AG, BF 1,481,786 (May 1967, France).

[13] Agfa-Gevaert AG, BF 1,479,891 (March 1967, France).

[14] Agfa-Gevaert AG, DBP 1,224,494 (Sept. 1966, Germany).

[15] Agfa-Gevaert AG, BP 1,108,612 (April 1968, England).

[16] H. Calvayrac and J. Golé, *Bull. Soc. Chim. France,* **1968**, 1076.

[17] (a) M. Kato, *Polymer Letters,* **7**, 605 (1969).
(b) M. Kato and M. Hasegawa, *Polymer Letters,* **8**, 263 (1970)

[18] (a) G. A. Delzenne and U. Laridon, *J. Polymer Sci.,* **22**, 1149 (1969).
(b) J. A. Arcesi, F. J. Rauner, and J. R. L. Williams: *2nd Retec Conf. on Photopolymers, Ellenville,* 1970, p. 19.

[19] M. Tsuda, *J. Polymer Sci.,* **7**, 259 (1969).

[20] F. Stuber, H. Ulrich, D. Rao, and A. Sayigh, *J. Appl. Polymer Sci,.* **13**, 2247 (1969).

[21] G. Wegner, *Makromol. Chem.,* **134**, 219 (1970).

[22] A. Hay, D. Bolon, K. Leimer, and R. Clark, *Polymer Letters,* **8**, 97 (1970).

[23] M. Tsuda, *J. Soc. Sci. Phot. Japan,* **32**, 51 (1969).

[24] R. B. Woodward and R. Hoffmann, *The Conservation of Orbital Symmetry,* Verlag Chemie, Weinheim, 1970.

[25] A. Reiser, G. Bowes, and R. J. Horne, *Trans. Faraday Soc.,* **62**, 3162 (1966).

[26] W. M. Moreau, *Polymer Preprints,* **10** (1), 362 (April 1969).

[27] R. Foster, D. Hammick, G. Hood, and A. Sanders, *J. Chem. Soc.*, **1956**, 4865.
[28] K. Nakamura and S. Kikuchi, *Bull. Chem. Soc. Japan,* **41**, 1977 (1968).
[29] M. Tsuda, *Bull. Chem. Soc. Japan,* **42**, 905 (1969).
[30] A. Reiser, H. M. Wagner, R. Marley, and G. Bowes, *Trans. Faraday Soc.*, **63**, 2403 (1967).
[31] A. Reiser and R. Marley, *Trans, Faraday Soc.*, **64**, 1806 (1968).
[32] A. Reiser, F. W. Willets, G. C. Terry, V. Williams, and R. Marley, *Trans. Faraday Soc.*, **65**, 3265 (1969).
[33] Gevaert-Agfa N.V., BPa 62709/69 (Dec. 1969, England).
[34] F. Lewis and W. Saunders, *J. Am. Chem. Soc.*, **90**, 7033 (1968).
[35] G. A. Delzenne, U. Laridon, and H. Peeters, unpublished results.
[36] R. M. Moriarty and R. Reardon, *Tetrahedron,* **26**, 1379 (1970).
[37] J. S. Swenton, T. J. Ikeler, and B. H. Williams, *J. Am. Chem. Soc.*, **92**, 3103 (1970).
[38] T. Tsunoda, T. Yamaoka, G. Nagamatsu, and M. Hirohashi, *2nd Retec Conf. on Photopolymers, Ellenville, 1970,* p. 25.
[39] H. Levine, L. Lesoine, and J. Offenbach, *Kodak Photoresist Seminar Proc.*, **1**, 15 (1968).
[40] A. G. Vavilov, *Zh. Nauch.i.Prikl.Fotogr.i.Kinematogr.*, **13**, 411 (1968).
[41] A. Mistr, M. Vavra, H. Adlerova and Z. Babak, *Collection Czech. Chem. Commun.*, **34**, 3811 (1969).
[42] U. Laridon and G. A. Delzenne, U.S. Pat. 3,427,161 (Feb. 69) Gevaert-Agfa N.V.
[43] M. Gilano and M. Lipson, *2nd Retec Conf. on Photopolymers, Ellenville, 1970,* p. 30.
[44] G. A. Delzenne, U. Laridon, and H. Peeters, *European Polymer J.*, **6**, 933 (1970).
[45] R. O. Symcox and J. D. Cotman, *J. Polymer Sci.*, **5**, 1165 (1967).
[46] Y. Kurusu, H. Nishiyama, and M. Okawara, *Kogyo Kagaku Zasshi,* **70**, 593 (1967).
[47] Y. Kurusu, H. Nishiyama, and M. Okawara, *Makromol. Chem.*, **133**, 49 (1970).
[48] C. H. Bamford, R. W. Dyson, and G. C. Eastmond, *J. Polymer Sci., Part C*, No. 16, 2425 (1967).
[49] G. Oster, Nature, **173**, 300 (1954).
[50] G. Smets, W. De Winter, and G. Delzenne, *J. Polymer Sci.*, **55**, 767 (1961).
[51] G. Manecke and G. Kossmehl, *Makromol. Chem.*, **70**, 112 (1964).
[52] G. Manecke and G. Kossmehl, *Makromol. Chem.*, **115**, 285 (1968).

[53] R. O. Lussow, *Photochem. Mach.–Photochem. Etch.* **4**, 2 (1969).
[54] R. K. Matta and L. C. Scala, *J. Electrochem. Soc.*, **115**, 247 (1968).
[55] R. Thornley and T. Sun, *J. Electrochem. Soc.*, **112**, 1151 (1965).
[56] B. Broyde, *J. Electrochem. Soc.*, **115**, 247 (1968).
[57] K. Kanaya, K. Shimizu, and U. Ishikawa, *Brit. J. Appl. Phys.*, **1**, 1657-65 (1968).
[58] L. Holland, K. Kanaya, K. Shimizu, and Y. Ishikawa, *Brit. J. Appl. Phys.*, **2**, 767 (1969).
[59] A. N. Wright, Report No. 67-C-290, General Electric (1967); US. Pat. 3,372,672 (March 1968) General Electric.
[60] J. F. Burgess and D. L. Schaeffer, *Photochem. Mach.–Photochem. Etch.* **4**, 2-5 (1969).
[61] General Electric, U.S. Pat. 3,489,563 (May 1967-Jan. 1970).
[62] General Electric, U.S. Pat. 3,346,384 (Apr. 25, 1963).
[63] Union Carbide, BPa 1,146,015 (June 66/March 69).
[64] M. Beisley, J. Castledine, and D. Cooper, *Electron. Letters* **5**, 257 (1969).
[65] J. R. Celeste and R. B. Heiart, *SPSE Proc., Cherry Hill, April 1970,* p. 42.

Organic Colorants for Polymers

THOMAS B. REEVE

Pigments Department
E. I. du Pont de Nemours & Co., Inc.
Wilmington, Delaware

ABSTRACT

Colorants vary widely in chemical structure with each type offering a combination of properties uniquely different in some respect from all others. Colorant behavior is also determined by external factors including the system in which it is incorporated and the conditions under which it is processed and used.

Information needed to assist the formulator in selecting organic colorants is outlined in this paper. This includes a definition of terms frequently used in describing colorants and their performance. Organic colorants are classified according to both color and chemical type and typical properties are described.

DEFINITION OF TERMS

An introduction to terms commonly used to describe colorants and their performance in polymers is needed to understand the discussion that follows.

Colorants are generally classified as either dyes or pigments. By definition, dyes are soluble in the system in which they are used, whereas pigments are essentially insoluble. For reasons that will become evident later, pigments are the principal colorants used in polymers of the type to be discussed here. A pigment is a very small particle-size solid which, because of its insolubility, must be uniformly and finely dispersed in the polymer by mechanical means to provide optimum properties. Essentially undiluted, high strength, dry organic pigments are known as toners. Organic colorants diluted with substantial quantities of dry colorless extenders such as calcium carbonate, barium sulfate, and metal resinate are frequently called lakes, although this term more strictly refers to chemical bonding or sorption of dyes on inorganic substrates. The term "extended" color is preferable. Water dispersible pigments are surfactant-treated dry forms that disperse readily in aqueous systems.

Most organic pigments pass through a press cake stage in the course of their manufacture. Aqueous dispersions are frequently made from press cake by adding surfactants and other agents to make them fluid, compatible, and stable in water systems. A flushed color is a pigment-in-oil (or resin) dispersion prepared by displacing water from the press cake without drying.

The polymer industry also makes frequent use of predispersed colorants, i.e., pigments that have already been dispersed in a medium or form that simplifies coloring of the finished product. Dry or paste concentrates are high loadings of pigment predispersed in resin binders that are compatible in the final system. Custom colorants are predispersed blends of pigments designed to match a desired shade; they may be obtained in either concentrate or final letdown levels.

The color characteristics of a pigment include a description of its opacity, relative tinting strength, and special optical effects (e.g., iridescence) in addition to its spectrophotometric response. Pigment cost normally refers to its price multiplied by the amount required to obtain desired color depth or opacity.

Dispersibility describes the ease with which pigment agglomerates can be reduced to an acceptable particle size. This is particularly important when polymers or dispersion equipment are involved in which it is difficult to develop high shearing stresses. Possible direct effects of inadequate dispersion include:

1. High cost – more pigment is required
2. Poor appearance – specking and streaking and low color intensity
3. Increased color control problems – hue variation with dispersion and color change during subsequent processing
4. Low mechanical strength
5. Poor film integrity – porosity and crocking

Weatherability refers to the ability to withstand all forms of weathering, including visible and ultraviolet light, moisture, temperature, and chemicals in whatever combination may exist during the life of the product. Weathering traits of importance include colorfastness, chalking, spewing, spotting, dirt collection, bronzing, cracking, checking, crazing, and retention of physical strength. Weatherability is often a function of the heat stability and chemical resistance of a pigment since these elements are encountered in the course of weathering.

Heat stability is defined as resistance to change in color or performance as a result of heat encountered during processing or in end product use. These changes may result from chemical decomposition, sublimation, crystal change, solution, or interaction with other ingredients in a compound. Heat stability is a function of temperature and time and may affect performance without affecting color (e.g., weathering and mechanical strength.)

Chemical resistance refers to inertness and compatibility with other ingredients within the compound as well as resistance to external environment. It is often associated with heat stability since reaction may take place during high temperature processing that would not otherwise occur.

Bleed/migration refers to bleed of a color through an overcoat of a different color, or migration of the color into an adjoining article. These are usually a result of slight solubility of pigment in the compound, although entrainment of pigment particles in other migrating materials can produce the same effect. Heat or presence of strong solvent materials in an adjoining surface will often accelerate the effect. Crocking describes the presence of dry color on a surface that can be rubbed off. It may be caused by bleed or migration, but may also result from poor dispersion, high pigment loading, or abrasion or deterioration of the surface such as through weathering. Bronzing is a form of crocking that produces an iridescent metallic luster on the surface.

Flocculation refers to the tendency for dispersed pigment particles to cluster together rather than remain uniformly dispersed. It occurs in systems which remain sufficiently fluid immediately after processing or application to permit movement of dispersed particles. Flooding is a phenomenon by which the various colorants in a system tend to separate into layers. Both situations result in inefficient use of pigment (e.g., loss of strength and opacity) and may affect performance as well (e.g., weatherability).

Crystal stability is of importance with those pigments that exist in more than one crystal form. Loss of strength or shift in hue may occur if crystal growth or phase change takes place. This is most likely to occur in systems containing strong solvents or during processing or storage at elevated temperature.

Plate-out describes the tendency of some pigments to adhere to the metal surfaces of processing equipment rather than remain in the system. This is often considered evidence of incompatibility in the system.

POLYMER INFLUENCE ON COLOR AND PERFORMANCE

The important role the polymer plays in colorant performance and selection must be recognized. It is evident from the properties described above that chemical reactivity within the system, as well as bleed and migration, are directly dependent on the polymer and other ingredients used. Flocculation and flooding tendencies are largely a result of interfacial surface tensions developed between the pigment and polymer, and heat stability requirements are a function of the temperature and conditions under which the polymer must be processed. Even weatherability is highly dependent on the polymer since each aids in the protection of the other, i.e., the colorant screens out ultraviolet radiation that may be harmful to the polymer while the polymer, through its ability to form a relatively impervious film, helps to protect the pigment from possibly injurious atmospheric conditions (e.g., chemicals and moisture).

Beyond these direct interdependencies, the wetting and working characteristics (e.g., rheology) of the polymer often have a major influence on the cost and quality of pigment dispersion that can be obtained, and the volatility and toxicity of the polymer influences the dispersion method used. Also, the inherent color, color permanence, transparency, and mechanical strength of the polymer will often influence colorant loadings used.

It is therefore apparent that polymer characteristics are an important consideration in the selection of colorants. The polymers this discussion covers are the conventional synthetic thermoplastic and thermoset resins and elastomers formed or applied by common molding, extrusion, and coating techniques. No attempt will be made to discuss color applications in traditional paint, ink, or natural fiber systems.

CHEMICAL CLASSIFICATION OF COLORANTS

Colorants vary widely in chemical structure, thereby accounting to a large degree for their wide range of colors and properties. Products of similar composition can be expected to behave alike; hence, classification according to chemical type and irrespective of color is useful in enabling us to predict performance.

Appendix I shows such a classification for organic pigments, including typical generic names and structures. Dyes are not included because of their very limited use in polymers of the type just described. Dyes of greatest interest are of the oil-soluble type, particularly azos and anthraquinones. They offer significant advantages over pigments of similar color in color intensity, transparency, strength, and economy; however, strong bleed and migration tendencies in many systems and generally poor heat stability and lightfastness often negates these advantages in favor of pigments. Principal dye applications are in rigid (i.e., unplasticized) thermoplastics like polystyrene, acrylic, cellulosic, rigid vinyl, polyester, and polycarbonate resins in which optical clarity is of great importance and likelihood of migration problems is small. Dyes, including the disperse type, are sometimes used for surface coloration of resins with the aid of alcohol and surfactant additives.

Pigments can first be classified into organic and inorganic types. Relative to inorganics, organic pigments are generally lower in specific gravity, higher in surface area, more transparent, stronger, higher priced (although not necessarily more expensive to use), more intense in color, less heat stable, and more soluble in organic solvents and polymers. However, there are many exceptions to these generalizations and the formulator would do well to consider at least the basic chemical types within each group for more reliable prediction of performance.

The azo pigment class is the largest of the organics in terms of both volume of consumption and variety of products available. Their colors are almost entirely in the yellow, orange, and red hues. They are characterized chemically by the presence of an azo chromophore (—N═N—) in the molecule which connects aromatic groups of varying complexity. Azo pigments may be further classified into azo dyestuffs, precipitated or metallized azos, and condensation azos. The important chemical difference between azo dyestuffs and precipitated azos is the absence in the dyestuffs of salt-forming carboxylic (—COOH) and sulfonic (—SO_3H) groups which tend to produce water solubility. Hence, this nonpolar nature of azo dyestuffs explains their tendency to be soluble in nonpolar organic vehicles (i.e., bleed) and resistant to acids and alkali. Conversely, the precipitated azos are less soluble and therefore more bleed resistant in organic media, but they are more likely to react with other ingredients in the system and with external chemicals with which they might come in contact.

To further understand differences in performance within the dyestuff group, it is helpful to divide these into mono and polyazo compounds. Molecular size increases with the number of azo units and this has the effect of increasing tinting strength, reducing solubility in organic vehicles (i.e., more bleed resistant) and improving heat and chemical resistance. The polyazos, the great majority of which are disazos (two azo groups), also tend to be more transparent and less lightfast in tints.

Poor heat stability and strong bleeding tendencies severely restrict use of monoazo dyestuffs in polymers of the type discussed here; the benzimidazolones are the best performers of this group but at considerably higher cost. The polyazos are generally more heat and bleed resistant, but the improvement is not sufficient to permit widespread use in these systems. Diarylide yellows are perhaps the most versatile, the OT type finding use in a number of high temperature applications (e.g., polyethylene wire coating), and the HR type having sufficient migration resistance for use in plasticized vinyls. Excellent chemical resistance makes the polyazos especially suitable for elastomeric and thermosetting systems where strong curing and oxidizing agents are often used. None of the polyazo dyestuffs is suitably lightfast in low concentration for long-term outdoor use.

The metallized azos are monoazos of varying chemical structure that are precipitated from their aqueous reaction medium in the form of a metal salt, metal chelate, or laked onto an adsorptive substrate, often in combination with a metal salt. The type of metal used, as well as the configuration of the coupling components attached to the azo group, account for the many color variations available in this group. Although more chemically reactive than azo dyestuffs, the range of bright hues, excellent bleed resistance, and low cost of these pigments has made them popular in a wide variety of polymers. Their sensitivity to chemicals is primarily a factor in rubber modified polymers (e.g., high impact polystyrene and ABS), polymers containing highly basic or acidic ingredients, or those containing ingredients which can enter into metal replacement (e.g., metal soaps and stabilizers). Attack by external chemicals of these types may also occur if the polymer is not sufficiently protective.

The metal salts and lakes are almost entirely reds and maroons. Heat stability varies from poor to good within the group, with Lake Red C and the calcium- and barium-precipitated Red 2B pigments being favored for high temperature applications. A number of the pigments in this group drift dark and blue when hot, but they return to their original color when cooled. Lightfastness is poor in low concentrations, but the Red 2B and BON types have found acceptance for noncritical outdoor applications in dark shades. The Red 2B pigments in particular tend to plate-out with other ingredients in improperly formulated vinyls. Also, metals used in the manufacture of precipitated azos can be injurious to some polymers.

The only chelated azo of importance is the nickel azo yellow. In contrast to other precipitated azos, this one offers excellent lightfastness in all concentrations, along with good heat stability and bleed resistance. Although chemical resistance is generally good, the relatively weak chelate bond makes the pigment subject to demetallization, which in turn degrades its other properties. This pigment is also considerably higher in cost than most other azo pigments currently available.

Condensation azo yellows and reds are high molecular weight disazo pigments that are structurally similar to conventional azo dyestuffs in that they contain no salt-forming groups or metal precipitating complexes that would introduce sensitivity to chemicals. At the same time, their higher molecular weight gives them generally better heat and bleed resistance than disazo dyestuffs. They are high strength, bright colors of relatively high cost. Tint lightfastness varies within the group, but it is decidedly better than most other azo pigments.

The phthalocyanines come the closest to representing the ultimate in pigment performance in the pigment consuming industries. Excellent fastness to heat, light, chemicals, and solvents is combined with high strength and transparency, low use cost, and intense shades. Their hue ranges from blue to green. The blue exists in at least two crystal forms, a red shade alpha form and the more stable green shade beta form. The green is derived by essentially complete chlorination of the blue; still yellower shade greens can be made by substituting bromine for chlorine. Most of the phthalocyanines available today are copper based, although metal-free products and those based on other metals are encountered on rare occasion. Although providing excellent performance in most applications, copper phthalocyanine blues have a tendency to drift greener and lighter in systems where slight solubility exists at high temperature (e.g., polystyrene and nylon), and crystal unstable forms are available which may lose color at moderately high temperatures or in solvent systems. Copper phthalocyanines can also act as nucleating agents in polyolefin polymers; this probably explains the warpage and shrinkage problems encountered in polyolefins when using phthalocyanines.

The quinacridones come the closest to duplicating the excellent performance of phthalocyanines in the orange, red, and violet portions of the spectrum. Good lightfastness in all concentrations, excellent bleed and chemical resistance, and adequate heat stability for processing in most resins has led to wide adoption of quinacridones in spite of their high cost. Performance in polystyrene, acrylic, and nylon resins is best in the lower temperature range because of limited heat stability in these resins. A number of the quinacridone products are difficult to disperse.

Vat dye pigments are derived from insoluble dyes containing two or more carbonyl ($>C{=}O$) groups. Vat dyeing refers to the process of applying the reduced leuco form of the dye to a textile fiber, after which the color is regenerated by oxidation. The vat dye pigment class includes those based on the thioindigo, anthraquinone, perinone, and perylene structures. These pigments cover a wide range of hues and are generally high strength, transparent, moderate to high cost, and intense in color. Fastness properties vary greatly, even within each chemical group; thus, it is difficult to generalize further on their performance. The thioindigos, as a group, have the poorest heat stability and both they and the anthraquinone types have a tendency to bleed and migrate. The anthraquinones are particularly susceptible to loss of color through chemical reduction. The lightfastness of most vat pigments is adequate in dark shades for long-term outdoor use, and some show excellent tint lightfastness as well.

Carbazole dioxazine violet is an exceptionally strong and intense violet that offers excellent lightfastness and bleed resistance in many systems. However, these properties can be degraded substantially by high temperature processing in some systems. This violet is commonly used for toning whites to overcome the slight yellowness of some systems and for shading phthalocyanine blue compositions to redder tones.

The isoindolinones are recent developments with commercial products available as yellows, oranges, and reds. They offer moderate to high color intensity, good lightfastness in many but not all systems, and generally good fastness to heat, chemicals, and solvents. Moderately high cost tends to limit their use, however.

Acid dye pigments are lakes of dyes containing sulfonic acid groups usually laked on alumina hydrate. Although high in strength and color intensity, poor lightfastness and bleed resistance severely limit their use. Water release from the hydrate can also be troublesome at high processing temperatures. Peacock blue is the best known acid dye pigment; it is used primarily in inks.

Basic dye pigments are basic dyes, characterized chemically by the presence of free or substituted amino groups in the molecule, which are converted into pigmentary form by precipitating with tannic acid, rosin or fatty acid soaps, phosphates, or such complex inorganic heteropoly acids as phosphotungstic and/or phosphomolybdic acids. Pigments in this group include the rhodamine reds, Victoria blue, Crystal and Methyl violets, and Malachite and Brilliant greens. Like the acid dye pigments, these offer high strength and intensity, but poor lightfastness and bleed resistance in water and solvents. Very little use is made of them in systems other than ink.

Pigment Green B, a complex iron salt of 1-nitroso-2-naphthol, is a strong, low cost, dull, olive shade of green in tint that is alkali and bleed resistant. However, tint lightfastness and heat stability are only fair and its iron content may be deleterious to some polymers. Its primary use is in rubber and alkaline emulsion coatings.

CLASSIFICATION ACCORDING TO COLOR

Useful as chemical classification may be for understanding and predicting performance, the usual problem facing a formulator is that the color to be matched has been predetermined, and he is next faced with selection of pigments that will produce that shade and offer the properties desired. Appendix II shows a classification of organic pigments according to color to aid in such selection. The more significant characteristics of the pigment that might influence selection are also shown.

TESTING OF COLORANTS

A great deal of significance is often attached to property descriptions such as those listed in Appendix II without due regard for conditions under which such performance can be expected and realistic consideration of the relative importance of the properties being sought. It is evident from the foregoing discussion that colorant performance is usually not absolute or intrinsic but depends on such outside influence as formulation (including colorant concentration), processing conditions, and circumstances of use. Product shape and dimensions, age and condition of ingredients, and uniformity of production introduce additional variables with which the formulator must reckon.

Ideally, a product should be tested under conditions identical to those of use. However, for reasons of time, cost, or difficulty in handling, simulated and accelerated testing is often used to predict results, and even this may be restricted to evaluation in "typical" formulas and only for investigating those few properties considered "critical." These compromises increase the possibility of inaccurate predictions of performance unless good correlation has been previously established. For example, accelerated weathering devices are frequently used to predict outdoor lightfastness of a colored product without recognizing that colorants often show different rates of failure under these exposure conditions than they would in actual use. Similar problems arise in other property tests, sometimes because failures are a result of a combination of factors rather than one alone. Accordingly, extreme care in interpretation of test results should be exercised.

CONCLUSION

Colorant properties are a function primarily of their chemical structure, but performance is influenced by the polymer formulation, processing conditions, and circumstances of use. Final decision on suitability of a colorant in a specific application should be made on the basis of appropriate testing for which good correlation with actual use has been established.

APPENDIX I – CHEMICAL CLASSIFICATION OF ORGANIC PIGMENTS

I. Azo Pigments

A. Monoazo Dyestuffs

Typical Names

Toluidine
Hansa*
Nitroaniline
Para
Arylide
Naphthol
Anisidine

Typical Structure

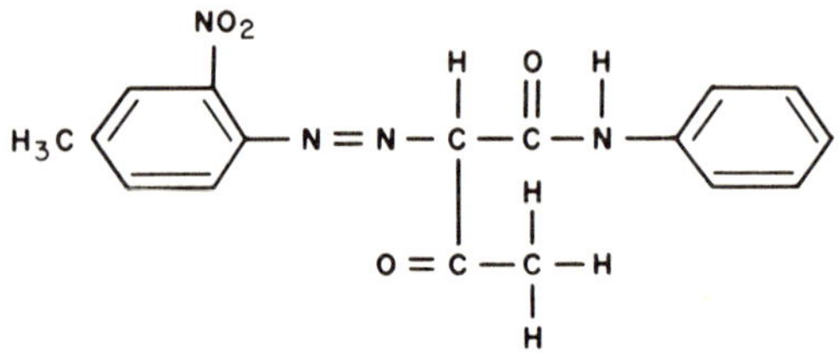

Toluidine Yellow (Hansa* G)
(Pigment Yellow 1, C.I. 11680)

B. Disazo Dyestuffs

Typical Names

Diarylide
Pyrazolone
Dianisidine

Typical Structure

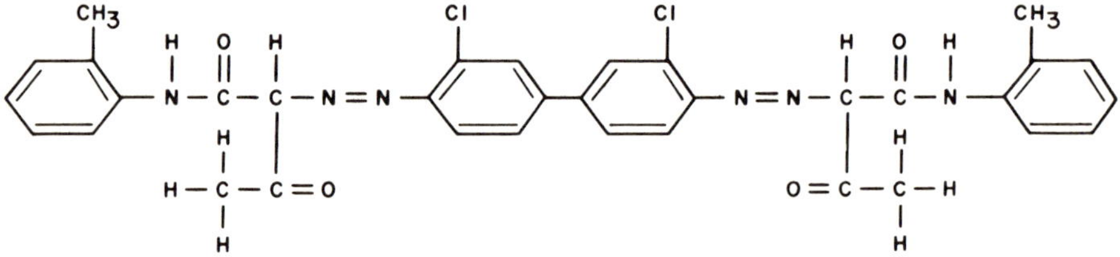

Diarylide Yellow OT
(Pigment Yellow 14, C.I. 21095)

*Hansa is a trademark of the GAF Corporation in the United States; however, the term is widely used in the description of monoazo yellows and oranges, including those of the basic structure shown here.

C. Precipitated Azos (Metal Salts or Lakes)

Typical Names

Lithol
Red 2G
Red 2B
BON
Pigment Scarlet
Lake Red C
Azo Bordeaux

Typical Structure

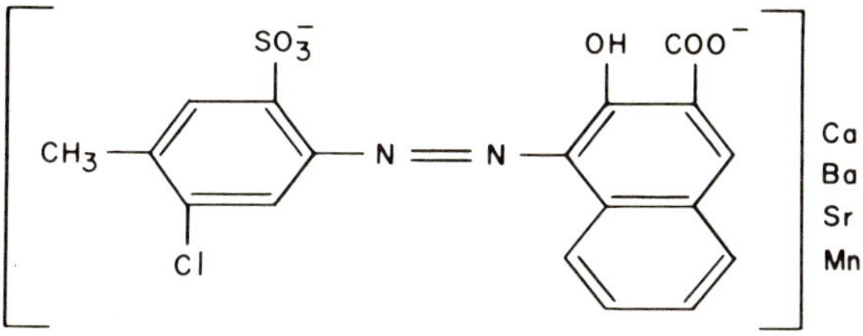

Permanent Red 2B
(Pigment Red 48, C.I. 15865)

D. Chelates

Typical Names

Nickel Azo
Para Brown

Typical Structure

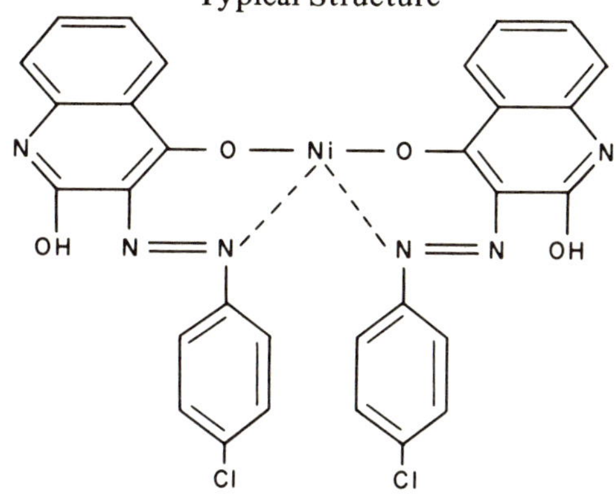

Nickel Azo Yellow
(Pigment Green 10, C.I. 12775)

II. Phthalocyanines

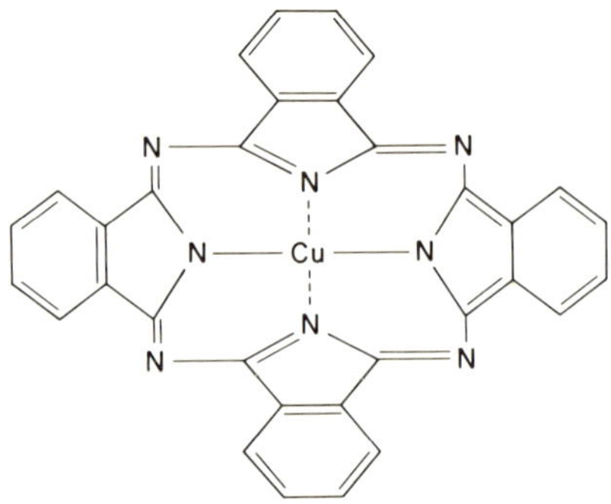

Copper Phthalocyanine Blue
(Pigment Blue 15, C.I. 74160)

III. Quinacridones

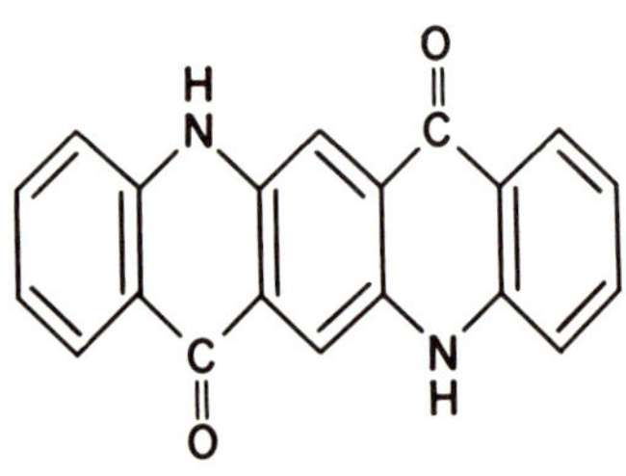

Quinacridone Red or Violet
(Pigment Violet 19, C.I. 46500)

IV. Vat Dye Pigments

A. Thioindigo

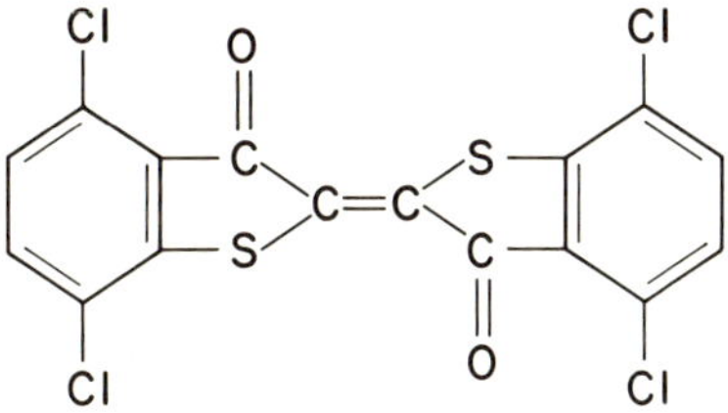

Thioindigo Red
(Pigment Red 88, C.I. 73312)

B. Anthraquinone

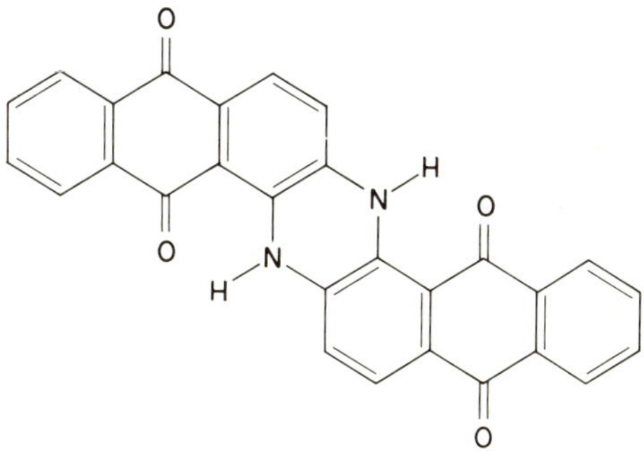

Indanthrone Blue
(Pigment Blue 64, C.I. 69825)

C. Perinone

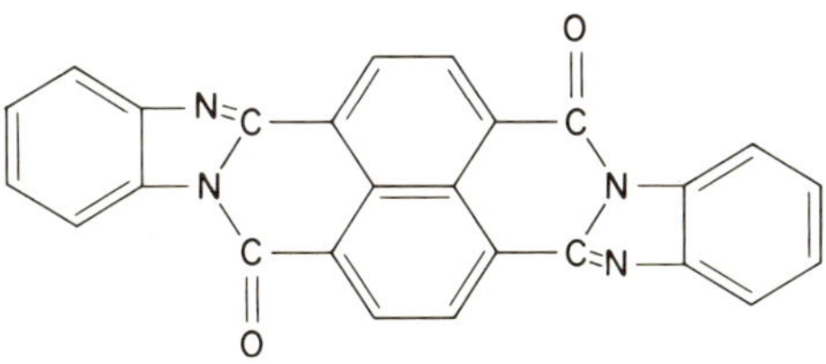

Orange GR
(Pigment Orange 43, C.I. 71105)

D. Perylene

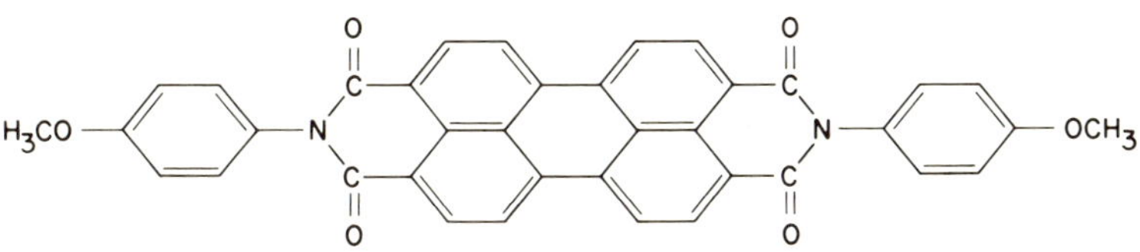

Perylene Red
(Pigment Red 190, C.I. 71140)

V. Dioxazines

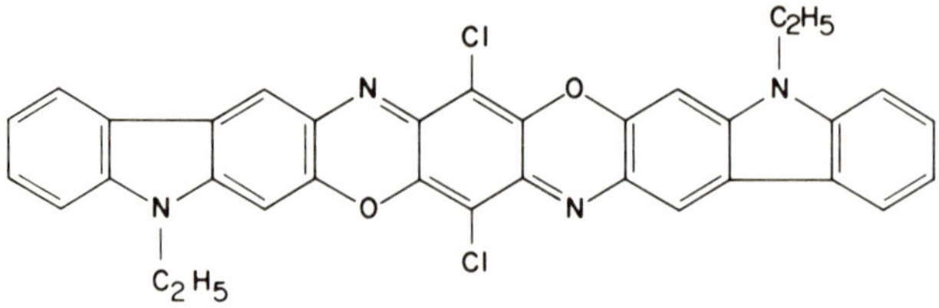

Carbazole Dioxazine Violet
(Pigment Violet 23, C.I. 51319)

VI. Isoindolinones

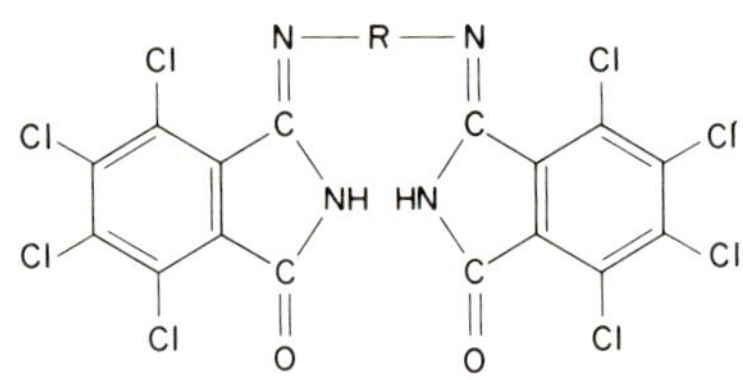

VII. Acid Dye Pigments

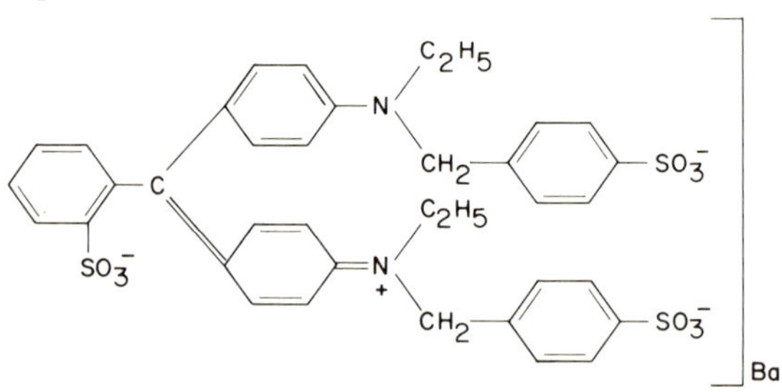

Peacock Blue
(Pigment Blue 24, C.I. 42090)

VIII. Basic Dye Pigments

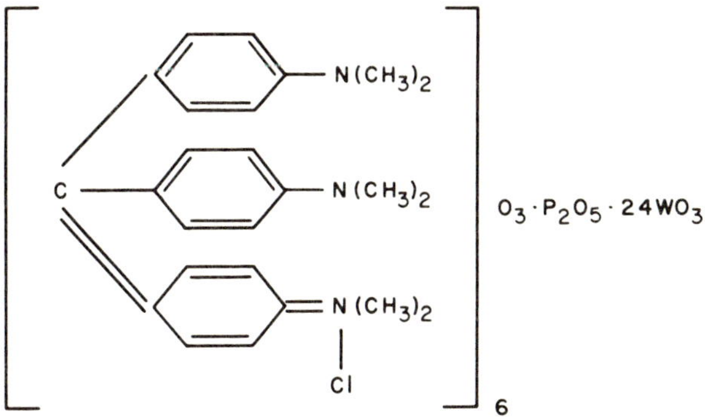

Crystal Violet
(Basic Violet 3, C.I. 42555)

APPENDIX II – TYPICAL PROPERTIES

A. Organic Yellows and Oranges for Polymers

Hansa* (monoazo)
Strong, intense colors; wide hue range. Bad bleed. Very poor heat stability. Good lightfastness (LF) in dark shades but poor in tint. Chemical resistant.

Diarylide (disazo)
Very strong, intense, transparent colors. Slight to moderate bleed. Fair to good heat stability. Very poor tint LF (except Yellow HR which is fair). High thixotropy in liquids. Excellent chemical resistance.

Nickel azo (chelated azo)
Greenish yellow, moderate strength and intensity, transparent. High cost. Good LF and heat stability. Subject to demetallizing with resultant bleed and poor LF.

Condensation azo
Strong, intense, transparent colors. Wide hue range. High cost. Fair to good tint LF. Good resistance to heat, bleed, and chemicals.

*See footnote, Appendix I.

Quinacridone gold and orange
Very transparent, dull tint. Good LF. Bleed and chemical resistant. Hard dispersing. High cost. Moderate heat stability.
Flavanthrone, Anthrapyrimidine (vat anthraquinones)
Strong, intense, transparent yellows. Very high cost. Good LF and chemical resistance. Moderate heat stability. Possible bleed. Subject to reduction.
Orange RK, Orange GR (vat anthraquinones)
Strong, intense colors. High cost. Good LF (especially RK) and chemical resistance. Moderate heat stability. Possible bleed (especially RK). Subject to reduction.
Isoindolinone
Strong, intense color. Moderate cost. Generally good LF and heat stability. Bleed resistant.

B. Organic Reds and Violets for Polymers

Para, Parachlor, Toluidine, Arylide (monoazo)
Strong, intense reds and maroons. Low cost. Bad bleed. Fair heat stability. Good LF in dark shades but poor in tint. Chemical resistant.
Pyrazolone (disazo)
Very strong, transparent, intense. Moderate cost. Slight to moderate bleed. Fair heat stability. Poor LF except in masstone. Chemical resistant.
Lithol (precipitated azo)
Strong, intense. Low cost. Bleed resistant. Poor heat stability and chemical resistance.
Permanent Red 2B, Pigment Scarlet, Lake Red C, BON, Lithol Rubine (precipitated azo)
Strong, intense reds and maroons; wide hue range. Low cost. Bleed resistant. Fair to good LF in dark shades but poor in tint. Fair to poor chemical resistance; subject to metal replacement. Good heat stability except in reactive systems; color drift when hot. Tendency to plate-out. Manganese type deleterious to some polymers.
Quinacridone
Strong, intense, transparent; wide hue range; LF in all concentrations. Bleed and chemical resistant (except to strong oxidizing agents). Generally good heat stability. High cost. Difficult dispersing.
Thioindigo (vat)
Strong, intense, transparent; wide hue range. Fair to good LF. Good

chemical resistance. Slight to moderate bleed. Fair to poor heat stability. Moderate cost.

Anthraquinone (vat)
: Strong, intense, transparent. High cost. Fair to good heat stability and LF. Chemical and bleed resistant; subject to reduction.

Perylene, Perinone (vat)
: Strong, intense, transparent. High cost. Good LF and heat stability in dark shades, fair in light. Chemical and bleed resistant.

Condensation azo
: Strong, moderately intense colors. Wide hue range. High cost. Fair to good tint LF. Good resistance to heat, bleed, and chemicals.

Isoindolinone
: Strong, moderately intense. Moderate cost. Fair heat stability. Good LF. Subject to bleed.

Carbazole Dioxazine Violet
: Strong, intense. Moderate cost. Fair heat stability. Good LF. Subject to bleed.

C. Organic Blues for Polymers

Phthalocyanine
: Strong, intense, transparent; hue range. Excellent LF and bleed and chemical resistance. Good heat stability; subject to phase change. Low cost.

Indanthrone (anthraquinone)
: Strong, moderately intense, red shade. Excellent LF. Fair heat stability. Good chemical resistance but subject to reduction. Possible bleed. Moderate cost.

D. Organic Greens for Polymers

Phthalocyanine
: Strong, intense, transparent color. Excellent resistance to heat, light, chemicals, and solvents. Moderate cost.

GENERAL REFERENCES

F. F. Ehrich, "Pigments, Organic," in *Kirk-Othmer: Encyclopedia of Chemical Technology*, Vol. 15, 2nd Ed., Wiley, New York, 1968, pp. 555-589.

H. Gaertner, "Modern Chemistry of Organic Pigments," *J. Oil Colour Chemists' Assoc.,* **46** (1), 13-46 (Jan. 1963).

A. P. Hopmeier, "Pigments," in *Encyclopedia of Polymer Science Tech.* (N. M. Bikales, ed.), Vol. 10, Wiley, New York, 1969, pp. 157-193.

E. E. Jaffe and W. J. Marshall, "Origin of Color in Organic Compounds," *Chemistry,* **37** (12), 6-13 (Dec. 1964).

H. A. Lubs, ed., "The Chemistry of Synthetic Dyes and Pigments," *American Chemical Society Monograph Series,* Reinhold, New York, 1955.

F. J. Nadolski, "Colorants for Rigid Thermoplastics," *Preprint, 21st Antec. of Soc. Plastics Engineers,* Vol. XI, 1965, pp. IX-2.

P. J. Papillo, "Mass Coloring of Plastics," *Mod. Plastics,* 44 (12), 131-188 (Dec. 1967).

L. S. Pratt, *The Chemistry and Physics of Organic Pigments,* Wiley, New York, 1947.

A. Pugin and J. von der Crone, "New Organic Pigments," *Official Digest, Federation Soc. Pt. Tech.*, **37** (488), 1071-1094 (Sept. 1965).

T. B. Reeve, "Pigmentation of Vinyls," in *Lecture Notes, PAG Seminar, Vinyl Plastics: Fundamentals of Materials and Compounding Practices,* Georgia Inst. Technol., Feb. 5-6, 1969.

T. B. Reeve, "Pigment Colors for Vinyl Coatings," *Color Eng.* **4** (4) 19-23 (July-Aug. 1966).

T. B. Reeve and W. L. Dills, "Principles of Pigment Dispersion in Plastics," *Preprint, 28th Antec of Soc. Plastics Engineers,* Vol. XVI, 1970, pp. 574-576.

J. Richter, "The Heat Stability of Some Recent Organic Pigments in Styrene Copolymers," *Preprint, SPE Retec, Coloring of Plastics-III,* New York, June 1966.

A. Siegel and W. S. Struve, "The Chemistry of Organic Pigments," in *Proceedings of the Perkin Centennial, 1855-1956* (H. J. White, ed), American Assoc. of Textile Chemists and Colorists, 1956, pp. 209-214.

J. E. Simpson, "Know Your Colorant," *Mod. Plastics,* 40 (3), 88-91, 199-210 (Mar. 1963).

J. E. Simpson, "Colorants for Plastics," *Mod. Plastics Encyclopedia, 1968-1969,* **45** (10A), 407 (Oct. 1968).

W. F. Spengeman, "Coloration of Plastics," *The Encyclopedia of Basic Materials for Plastics,* Reinhold, New York, 1967, pp. 125-129.

W. F. Spengeman and G. Wormald, "Lightfastness Testing of Pigment Colors," *ASTM Bull.,* No. 249 (Oct. 1960).

W. S. Stoy, "Colorants," in *Mod. Plastics Encyclopedia, 1970-1971,* **47**(10A), 326 (Oct. 1970).

D. G. C. Thornley, "Pigments, Plastics and Progress," *J. Soc. Dyers Colourists,* 86 (1), 13-19 (Jan. 1970).

G. Wormald and W. F. Spengeman, "Pigment Colors for Plasticized Vinyl

Chloride Polymers," *Ind. Eng. Chem.*, **44**, 1104-1107 (1952).

R. H. Zabel, "Colorants for Plastics," *Mod. Plastics Encyclopedia, 1969-1970,* **46** (10A), 266 (Oct. 1969).

Author Index

Numbers in parentheses are reference numbers and indicate that an author's work is referred to although his name is not cited in the text. Numbers in italics give the page on which the complete reference is listed.

C

G

H

P

Q

R

S

Subject Index